T0328381

# Innovation Matters

# Innovation Matters

## Competition Policy for the High-Technology Economy

Richard J. Gilbert

The MIT Press
Cambridge, Massachusetts
London, England

The open access edition of this book was made possible by generous funding from Arcadia—a charitable fund of Lisbet Rausing and Peter Baldwin.

This book was set in Palatino by Westchester Publishing Services.

Library of Congress Cataloging-in-Publication Data

Names: Gilbert, Richard J., 1945- author.
Title: Innovation matters : competition policy for the high-technology economy / Richard J. Gilbert.
Description: Cambridge, Massachusetts : MIT Press, [2020] | Includes bibliographical references and index.
Identifiers: LCCN 2019039525 | ISBN 9780262044042 (hardcover), 9780262545792 (pb)
Subjects: LCSH: High technology industries. | Competition. | Antitrust law--Economic aspects. | Consolidation and merger of corporations--Law and legislation--Economic aspects.
Classification: LCC HC79.H53 G56 2020 | DDC 338.8/2--dc23
LC record available at https://lccn.loc.gov/2019039525

To Michael and (Esther)$^2$

# Contents

# 1    Introduction

Current antitrust enforcement has its priorities backwards... the promotion of production and innovation efficiency should be the first economic goal of antitrust policy.
—Joseph F. Brodley, "The Economic Goals of Antitrust: Efficiency, Consumer Welfare, and Technological Progress" (1987)

There is widespread concern that competition is not working for the high-technology economy. Dominant firms supply many information, computing, and internet services. Business creation has slowed. Venture capitalists shun start-ups that would compete with the major digital platforms. Firms that dominate high-tech industries have managed to acquire or eliminate many potential competitors. Politicians on the left and the right are demanding more aggressive antitrust enforcement, including breaking up high-tech titans or limiting their operations to prevent them from discriminating against firms that rely on their services.

Some lay the blame for the failure of competition in the high-technology economy on misguided antitrust enforcement, which has been captured by economic arguments that equate consumer welfare with low market prices. There are calls to replace this consumer welfare standard with alternative approaches that allow a broader consideration of the effects of policies on dimensions other than price, such as jobs, privacy, inequality, and the concentration of political power.[1]

A focus on consumer welfare has been a stabilizing influence for antitrust enforcement. Alternative goals are often less precise or they admit policies that do not benefit consumers in the near term or in the more distant future. The main thesis of this book is that antitrust enforcement has to change to address challenges to competition in the high-technology economy, and that positive change can occur without sacrificing a focus on consumer welfare. The answer is to move from *price-centric* to *innovation-centric* competition policies. The transition will require a different emphasis in antitrust enforcement, different analytical approaches, and substantive changes to methodologies and presumptions that have been adopted by the courts that enforce the antitrust laws.

Antitrust agencies have taken steps to address innovation in their enforcement decisions. In 1993, General Motors (GM) proposed to sell its Allison Transmission Division to ZF Friedrichshafen AG. The companies were two of the world's largest manufacturers of automatic transmissions for large trucks, buses, and other commercial and military vehicles. At the time, I was the Deputy Assistant Attorney General for Economics in the Antitrust Division at the US Department of Justice (DOJ).[2] There were serious concerns in the Division about the transaction, but they did not fit into the usual enforcement boxes. Although Allison and ZF competed in Europe, ZF was only a minor manufacturer of medium- and heavy-duty automatic transmissions for vehicles sold in the US. Because ZF had a small share of sales in the US, the transaction was unlikely to have a very large effect on the prices paid by US customers.

But lawyers and economists in the Antitrust Division had a different concern: The merger (technically, an acquisition of GM's Allison Division by ZF),[3] if allowed to occur, would have reduced the merged company's motivation to innovate by eliminating rivalry between Allison and ZF in Europe, which would have had negative consequences for buyers in the US. (By "innovation," I mean a new or improved product or process that differs significantly from previous products or processes.[4] Innovation is more than invention, which is the act of discovering a new product or process, because innovation requires that an invention be put into active use or be made available for use by others.[5]) The GM-ZF transaction, if allowed to occur, would have denied US buyers the benefits of new or improved products that both GM's Allison Division and ZF would have had an incentive to develop if they remained independent.

The DOJ challenged the proposed merger.[6] The complaint emphasized harm to innovation as well as traditional adverse price effects. In its complaint, the Antitrust Division defined an "innovation market," which included Allison and ZF as two of the most important competitors engaged in research and development (R&D) for automatic transmissions used in large trucks, buses, and other commercial and military vehicles, and alleged that the merger would create a near-monopoly in the innovation market and reduce incentives to innovate. The parties abandoned their proposed merger in response to the complaint.

The GM-ZF merger challenge caused a stir in the staid antitrust community. Some accused the Antitrust Division of ignoring accepted antitrust principles by losing its traditional focus on price effects. Others railed that too little was known about innovation incentives to admit such concerns into the antitrust policy sphere. Despite these objections, the merger protected innovation incentives for automatic transmissions used in large trucks and buses, which is what we would expect today from developments in economic theory and empirical research on market structure and innovation.

Since the GM-ZF complaint, there have been more calls for antitrust enforcement to challenge mergers and other conduct that would harm innovation. In the late 1990s, the Antitrust Division investigated whether Microsoft's conduct related to its Windows personal computer (PC) operating system and Internet Explorer browsing software promoted innovation or was an abuse of monopoly.[7] In 2001, the Federal Trade Commission (FTC) investigated whether a merger of Genzyme and Novazyme would promote or delay a cure for a genetic disorder that is often fatal for hundreds of young patients. More recently, the FTC had to decide whether changes to Google's search engine were a consumer-friendly innovation or an anticompetitive design change that excluded competitors. Since the GM-ZF case, both the Antitrust Division and the FTC have addressed innovation concerns in numerous other proposed mergers, acquisitions, and joint ventures.

Antitrust enforcers often voice their concerns about protecting innovation, and the history of antitrust legislation supports the objective of preserving opportunities for dynamic competition. Nonetheless, antitrust enforcement evolved over more than a century to promote price competition by preventing mergers or other conduct that may widen the gap between prices and production costs, often to the exclusion of concerns about innovation.

The foundational statutes for US antitrust law are the Sherman and Clayton Acts. Section 1 of the Sherman Act prohibits contracts, combinations, or conspiracies in restraint of trade.[8] Section 2 does the same for monopolization or attempts to monopolize.[9] Section 7 of the Clayton Act prohibits mergers and acquisitions whose effects may be substantially to lessen competition, or to tend to create a monopoly.[10] The FTC addresses unfair and deceptive methods of competition under Section 5 of the Federal Trade Commission Act.[11] The statute covers conduct that would violate the Sherman Antitrust Act or the Clayton Act, but may address other conduct that is likely to harm competition.[12]

Other jurisdictions have their own versions of the Sherman and Clayton Acts.[13] For instance, Article 101 of the Treaty on the Functioning of the European Union prohibits cartels and other agreements that could disrupt free competition in the internal market of the European Economic Area (EEA), and Article 102 prohibits any undertaking that holds a dominant position in a market from abusing that position. Council Regulation 139/2004 prohibits mergers and acquisitions that would significantly reduce competition in the EEA.

It is understandable why antitrust enforcement came to focus its power on price competition. The proscriptions in the antitrust statutes are vague, and courts have had to create their own guidebooks to interpret them. Economic theory provided an internally consistent description of how price competition benefits consumers and established a nexus with the apparent objectives of the antitrust laws. Courts, antitrust agencies, and economists developed tools that facilitate a quantitative evaluation of price impacts. The promotion of price competition was not necessarily the only objective of antitrust law, but the application of economic methodologies to evaluate and promote price competition was something that the courts could do well.

This book is an attempt to collect in one place the current state of knowledge about antitrust enforcement for innovation and price competition for future products and services, to complement the state of knowledge about antitrust enforcement for price competition for existing products and services. The narrative is directed to an audience of economists, competition enforcers, and practitioners, although I hope that the book will appeal to others with interests in competition policy.

Chapter 2 describes the distinctive features of the high-tech economy and the challenges they raise for antitrust enforcement. These features include the potential for industry disruption, network effects, the importance of intellectual property, and the fact that many high-tech firms operate as platforms that coordinate prices and terms of service for different firms and users. Network effects reinforce the dominance of major internet companies because consumers value the participation of other consumers in their services. Intellectual property, economies of scale from R&D and the aggregation of data, and platform characteristics such as zero prices for some services, erect additional barriers to new competition and complicate the evaluation of alleged antitrust harms. Furthermore, tech titans have developed reputations for acquiring potential competitors and for competing aggressively against start-ups that attempt to enter markets that they serve, or that they have the capabilities to serve. Chapter 2 describes the frenetic pace of acquisitions by Google's parent company Alphabet, Facebook, Apple, Amazon, and Microsoft.

This chapter addresses whether the distinctive features of the high-technology economy warrant a different or more aggressive approach to antitrust enforcement. The antitrust laws are sufficiently flexible to allow for innovation-centric competition policy; however, the laws have been interpreted over time in ways that raise obstacles to sound enforcement policies for innovation. Many legal precedents have evolved that are not helpful for an evaluation of harm to innovation. These precedents support measures that promote short-run economic efficiency by moving prices closer to marginal production costs. While this evolution has had positive results for consumer-friendly competition enforcement in "old economy" industries such as manufacturing, mining, and services, it does not necessarily promote innovation, which requires the expectation of positive profits to motivate investment in R&D.

A major obstacle to an innovation-centric competition policy is the traditional emphasis in antitrust litigation on market definition and market shares. Market definition identifies the products and services that are relevant to an antitrust evaluation and their geographic locations. Firm market shares follow from the calculation of sales, revenues, or other relevant firm characteristics, such as production capacities, in the defined markets. Market definition and the calculation of market shares often are not useful analytical tools for a merger or conduct by a firm that is likely to harm incentives to invest in R&D or threaten competition in a future market. The precise boundaries of a market that does not presently exist are inherently uncertain. Moreover, given relevant future markets, available data at best allows a prediction of firm shares in these future markets.

In some other respects, traditional price-centric antitrust enforcement policies do not conflict with policies that prevent harm to innovation. For example, as discussed in detail in chapter 8, courts applied conventional antitrust principles to evaluate allegations that agreements between Microsoft and suppliers of software and hardware excluded competition in violation of Sections 1 and 2 of the Sherman Act. The resulting enforcement outcomes generally aligned with policies that are more focused on innovation, although the ability of network effects to reinforce market dominance calls for stiffer enforcement of exclusionary agreements. Conduct that makes it more difficult to attract customers to a competing product can cause a market with network effects to favor a dominant supplier, even though the conduct falls short of the substantial foreclosure standard that most courts have adopted for unlawful exclusive dealing under Section 1 of the Sherman Act.

Chapter 3 focuses on two fundamental themes of innovation competition: the "replacement effect," first described by Kenneth Arrow,[14] and the "Schumpeterian" theory of imperfect competition and the appropriation of private returns for R&D.[15] These two themes have dramatically different implications. Kenneth Arrow pointed out that the existing profits that firms earn in imperfectly competitive markets can dull innovation incentives. The incentive to innovate is the difference in a firm's profit with and without an innovation. This difference is reduced if an innovation replaces profits that firms earn from their existing products or technologies. In contrast, Joseph Schumpeter argued that imperfectly competitive markets provide innovation incentives that are absent in highly competitive markets by making it easier for firms to profit from their discoveries and by providing a more stable flow of earnings to cover the costs of R&D.

Both themes provide valuable insights, but they omit important considerations that can change their predictions. Arrow explained the replacement effect in a highly simplified model that abstracts from R&D competition and industry dynamics and is limited to process innovations that lower a firm's production cost. For example, contrary to Arrow's prediction that monopoly power discourages incentives to invest in R&D, monopoly power can have the opposite effect if innovation allows the firm to maintain its monopoly by preempting competition from potential rivals. Incentives for product innovations are more complex than incentives for process innovations because a firm can benefit by coordinating prices for existing and new products.

Modern economic theory and empirical evidence suggest that industry concentration can allow innovators to appropriate greater profit from their innovations under some circumstances. This is consistent with Schumpeter's argument that imperfectly competitive markets promote innovation incentives, but empirical evidence does not generally support a Schumpeterian perspective that monopoly power promotes innovation. There is no evidence that monopoly encourages R&D investment by providing a more stable flow of earnings. Some (but not all) empirical studies discussed in chapter 6 show greater R&D investment or innovation output in more competitive markets, and empirical studies do not generally support a conclusion that mergers promote R&D investment or innovation.

Chapter 4 provides further elaboration of the complex interactions between competition and innovation incentives. This chapter addresses issues that include market dynamics, cumulative innovation in which discoveries build on prior discoveries, and managerial and organizational theories of corporate behavior. Simple models of innovation competition and races to patent a discovery generally show that an increase in rivalry increases the probability of discovery and advances the likely date of a discovery. More complicated dynamic models capture the interdependence between market structures that motivate investment in R&D and the market structures that result from successful innovation. These theories show that competition can reduce the rate of innovation in some instances and demonstrate the importance of technological differences among firms for innovation incentives. Theories of corporate behavior for innovation emphasize cognitive distortions and organizational adjustments that cause dominant firms to ignore or eschew innovation opportunities, although predictions of the theories often are not fundamentally different from predictions of models that focus solely on economic incentives.

Antitrust authorities and the courts have limited policy levers to influence innovation. Antitrust enforcement can restrain single-firm conduct, establish limits on permissible agreements, and either prevent mergers and acquisitions or condition them on structural or behavioral remedies. Neither the antitrust authorities nor the courts can control competition directly. Chapter 5 addresses theoretical issues that are relevant to the analysis of the effects of mergers on innovation incentives and future price competition. In recent years, almost every challenge by US antitrust authorities of a merger or acquisition in a high-tech industry has included an allegation of harm to innovation. Yet the Horizontal Merger Guidelines, a joint publication of the DOJ and the FTC, barely mentioned innovation until the most recent revision of these guidelines in 2010.

The major tech platforms are adept at identifying potential competitors and acquiring them before they can achieve a scale that triggers antitrust review. The ability of many high-tech firms to identify and acquire promising competitors justifies greater antitrust scrutiny of acquisitions of potential competitors. Courts have been reluctant to challenge the acquisition of a potential competitor absent clear evidence that the potential competitor would have entered the relevant market without the acquisition. A recommendation offered in this chapter is that courts should reverse this historical reluctance if the potential competitor is an innovator. If the acquisition of a successful innovator would harm competition, antitrust enforcers should block the acquisition even if the probability of success is small, unless the acquisition has other efficiency benefits. A qualification is that, in some instances, the opportunity to sell a start-up or promising R&D project to an established firm is the most powerful incentive for innovation and the best way to commercialize a new product. Prohibiting acquisition by an established firm could discourage innovation if the innovator cannot partner with the acquiring firm for which it offers the most value. Furthermore, some established companies are likely to compete directly against start-ups if they cannot acquire them, and the threat of this competition can be a significant deterrent for innovation by new entrants.

Chapter 6 reviews the empirical literature related to competition, mergers, and innovation. The empirical evidence for a link between competition and innovation is somewhat mixed. While several studies show that competition promotes innovation, others find either a negative effect or no effect. One result that appears in several empirical studies is that the positive effect of competition on innovation is greater for firms that are at or near the frontier of efficient production. Competition has a less-positive effect, and may discourage innovation, for firms that substantially lag their rivals. These empirical results are consistent with the theory described by the Arrow replacement effect and Schumpeterian appropriation incentives.

A merger differs from a reduction in competition because it leaves the R&D assets of the merging firms intact, at least in the near term, but centralizes control of the merging parties' R&D decisions. Because mergers are related to, but not equivalent to, a reduction in competition, this chapter summarizes the empirical literature on competition and mergers separately.

Only a few empirical studies apply sophisticated statistical techniques to uncover the effects of mergers on R&D and innovation, and the few sophisticated studies do not identify a consistent pattern of results. Furthermore, observations are censored because antitrust authorities challenge mergers that they believe have anticompetitive effects, and consequently these mergers would not appear in the data. Despite these limitations, these studies do not support a conclusion that mergers generally promote R&D investment or innovation.

Case studies are useful illustrations of the successes and failures of antitrust policy for innovation. I begin in chapter 7 with examples of merger enforcement by US and European antitrust agencies in cases that alleged innovation concerns. The chapter reviews several instances in which the agencies refused to accept a structural or behavioral remedy to address their competition concerns and consequently the parties abandoned the proposed transactions. The challenges reviewed in this chapter appeared to have restored innovation incentives and future price competition that the agencies alleged would have been harmed by the merger.

In most proposed mergers and acquisitions, the antitrust agencies resolve their innovation concerns by negotiating consent decrees that mandate partial divestitures or licensing agreements. Chapter 7 reviews several of these consent decrees and follows the performance of entities that were the recipients of divested assets or patent licenses. Some of the divestiture agreements appear to have achieved the objective of restoring innovation incentives that might have been lessened by the proposed transaction, while others appear to have been less successful. For some proposed mergers or acquisitions in which the parties agreed to divest R&D assets to a third party, there is little evidence that the recipient of the divested assets continued to invest in R&D directed toward the applications for which the antitrust agency expressed innovation concerns. Broad licensing obligations have had a better success record. The merging parties and the industry as a whole continued to invest in R&D and file for patents at rates that were comparable to or higher than the premerger levels.

Chapters 8 and 9 deal with single-firm conduct that affects innovation by examining and inferring policy lessons from two significant examples. Chapter 8 discusses the antitrust case brought by the US DOJ and several states against Microsoft for monopolizing the market for PC operating systems. This chapter also describes cases brought by the European Commission (EC) that challenged Microsoft's conduct related to media players and workgroup servers.

The Microsoft cases illustrate several themes that are explored throughout this book. In the US case, the appellate court recognized the challenges of crafting appropriate antitrust enforcement for a dynamic market characterized by strong network effects, but it rejected the argument that the antitrust laws are not applicable to firms that operate in the high-technology economy. The court largely applied traditional antitrust principles, but it also carved out differential treatment for linking software products when it refused to condemn the tying of the Internet Explorer web browser to the Windows operating system.

A central allegation in the US case and in a related case brought by the EC was that Microsoft's actions prevented the Netscape internet browser from undermining Microsoft's monopoly by becoming a platform to develop applications that would run on different operating systems. That has yet to occur. Nonetheless, the consent decree that ended the US litigation and decisions by the EC had beneficial effects for software innovation by constraining conduct by Microsoft that would exclude competition and by encouraging Microsoft to make its software products interoperable with other products.

Chapter 9 describes investigations by the FTC and the EC that addressed the display of Google search results for comparison shopping services (CSS). CSS websites collect product offers from online retailers and allow users to click on links to the retailers' websites to make a purchase. A redesign of Google's search algorithms gave prominent position to its proprietary CSS in response to relevant queries, while demoting independent CSS websites in Google search results. The redesign caused a substantial reduction of internet traffic to independent CSS websites, and concentrated consumer attention on Google's proprietary CSS. Although the Microsoft cases largely addressed the company's efforts to maintain its monopoly power in PC operating systems, the Google case addressed conduct that arguably extended that company's monopoly power in internet search and advertising to the related activity of comparison shopping services.

The Google Shopping case offers insights into the antitrust treatment of broad categories of conduct that arise in the context of other high-technology platforms. One category is the incentive and ability of digital platforms to preference their products and services over those of their rivals. The alleged preference in the Google Shopping case is Google's demotion of rival CSS websites in its search results and the prominent placement of its own CSS in response to product queries. Similar allegations have been raised about preferential placement by Amazon for its private label products on its online retail platform and allegations that Apple favors its proprietary apps in response to app searches.

A second broad category is the antitrust treatment of innovations and product designs that exclude rivals without compensating consumer benefits. Concerns about the ability of dominant high-tech firms to design their products in ways that imitate and eliminate potential competition have had repercussions for innovation by potential rivals. Venture capitalists describe a "kill zone" of technologies that surround the businesses of the tech titans. Technologies in the kill zone are unattractive for venture capital because there is a high risk that the dominant firm will extinguish independent innovators if they are successful. The Google Shopping case illustrates the challenge of identifying and enforcing anticompetitive conduct when dominant firms can easily integrate into related operations.

The FTC and the EC focused on similar issues in the Google Shopping case and studied similar evidence, but their investigations had different outcomes. The FTC decided not to challenge Google's conduct related to the design of its search displays; the EC fined Google for violating European antitrust law and ordered the company to design a search display that does not preference its own CSS. The disparate outcomes reflect different approaches to product designs that can exclude competition in the two jurisdictions. Unfortunately, the FTC did not explain the reason for its decision any detail. The EC published a detailed decision, but it did not explain how it evaluated the costs and benefits from Google's conduct.

This chapter describes several tests that have been proposed to identify conduct related to innovation and product designs that have anticompetitive effects (sometimes called "predatory innovation"). Each of these tests has significant limitations. The chapter concludes that the most useful analytical approach is a truncated rule of reason that exempts substantial new designs or innovations from potential antitrust liability unless they are accompanied by other conduct that has exclusionary effects without compensating benefits. Product designs or claimed innovations that have little or no merit would be candidates for a full rule of reason analysis that evaluates the benefits from these designs or claimed innovations and compares them to their exclusionary effects.

Following the truncated rule of reason, Google's conduct in the comparison shopping case would escape antitrust condemnation if a court concludes that its proprietary shopping product is a significant innovation and if Google has a pro-competitive justification for demoting competing CSS websites in its search results. This approach would not exonerate Google if the demotion of competing CSS websites has little or no efficiency justification.

Next, chapter 10 addresses antitrust policy for standards and for conduct that affects interoperability or compatibility. Two or more systems are interoperable if they can communicate efficiently with each other. Applications are compatible if they can function within the same work environment, such as Microsoft Word and Excel. Interoperability is sufficient but not necessary for compatibility. Interoperability standards can promote innovation by allowing firms to specialize in components and exploit economies of scale with the knowledge that their components will be compatible with other components that together provide valuable services. However, standards also have antitrust risks. Dominant firms can exclude rivals by unilaterally promoting a standard that is not compatible with products supplied by their rivals. Cooperative standard-setting raises the types of risks that are common when actual or potential competitors discuss their joint interests in commercial applications. Intellectual property (IP) rights further complicate the standard-setting process because standards can confer substantial market power on owners of IP rights that are essential to make, use, or sell products that comply with the standards.

Finally, chapter 11 concludes with some remarks regarding the adaptations that courts and antitrust authorities must make to implement innovation-centric competition policy. The chapter closes with comments about the suitability of structural reforms to address the ability and incentives of major tech platforms to harm competition and innovation.

# 2 Should Competition Policy Differ for the High-Technology Economy?

An antitrust policy that reduced prices by 5 percent today at the expense of reducing by 1 percent the annual rate at which innovation lowers the cost of production would be a calamity. In the long run, a continuous rate of change, compounded, swamps static losses.
—Judge Frank Easterbrook, "Ignorance and Antitrust" (1992)

## 1  Innovation in the US Economy

Private businesses are the engines of technological progress in the US. About 70 percent of US research and development (R&D) spending occurs at private businesses, with the remainder split between higher education, nonprofit organizations, and federal entities. In the aggregate, US industries spent about 4 percent of their revenue on R&D in 2015. Some firms spend nothing on R&D. Many firms in high-tech industries such as computers and pharmaceuticals spend 10 percent or more of their revenue on R&D (figure 2.1).

Industries with high R&D intensities (defined by R&D expenditure as a percentage of total revenue) comprise the core of the high-technology economy. In addition to pharmaceuticals and electronic components, other R&D-intensive industries include communications equipment, software, information and internet services, data processing, aerospace products, scientific instruments, chemicals, and scientific services. For firms in the high-technology economy, it is critical that competition policy consider likely effects on innovation incentives, in addition to the traditional policy focus on conduct that may raise prices.

Invention and innovation have many determinants that are beyond the purview of competition policy. Inventions often spring from the spark of genius by a single entrepreneur driven by a desire to solve a

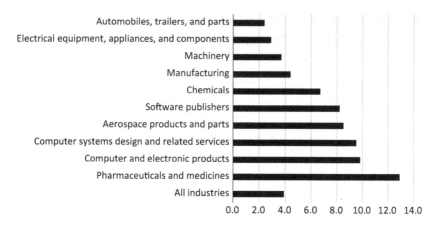

**Figure 2.1**
R&D expenditures as a percent of revenues for selected US industries (2015).
Source: National Science Foundation (2018a) and (2018b).

problem or create a new product without paying particular attention
to monetary rewards. Others, such as participants in the open-source
software community, collaborate to share knowledge and create new
business models that are not predicated on commercial success.[1] But
most applied innovations in the modern economy result from industry
teams driven in large part by expected profitability.[2] Competition policies can shape economic incentives to invest in innovative activities by
preventing mergers or other conduct that would harm innovation
incentives.

Antitrust scholars and practitioners debate whether characteristics
of high-tech firms justify more or less antitrust enforcement, or perhaps
an entirely different approach to competition policy for the high-
technology economy.[3] Section 2 of this chapter describes distinctive
characteristics of high-tech industries. Section 3 asks whether traditional antitrust enforcement is up to the task of preventing conduct,
such as acquisitions by dominant firms of potential competitors, that
harms innovation. There is general agreement that the statutes that
govern antitrust enforcement in the US and other developed economies
are flexible enough to account for the complexities of high-tech markets.
However, courts and enforcement agencies have applied antitrust laws
with the objective of preventing harm to competition for existing products, while paying less attention to factors that may harm innovation
or competition for new products. Section 4 demonstrates that moving
from a price-centric to an innovation-centric focus for competition

policy will require significant changes in the way that courts evaluate competition. Some of these requirements appear again in the concluding chapter 11 of this book, but they are sufficiently important for innovation-centric competition policy to warrant repetition. It remains to be seen whether courts will rise to this challenge.

## 2   The High-Technology Economy Is Different

The high-technology economy has many characteristics that differentiate firms and industries in this sector. The distinctions are a matter of degree. These characteristics are not necessarily absent from firms and industries that do not have high R&D intensity, and not all industries with high R&D intensity possess all these characteristics. Furthermore, the list of high-tech firms and industries is ever-changing. Amazon began as a low-tech online bookseller before it distinguished its platform with advanced logistics and moved into other services such as cloud computing. For many years, motor vehicle manufacturers invested approximately the economywide average share of their revenues in R&D, but that may increase as motor vehicles incorporate more self-driving features.[4]

### Innovation can disrupt industries and complicate predicted effects of antitrust enforcement

The high-technology economy is dynamic, with waves of innovation that can destroy incumbents. The dramatic growth of companies such as Microsoft, Amazon, Alphabet (the corporate parent of Google), and Facebook is testament to the power of "creative destruction," a term coined by Joseph Schumpeter in 1942. He described a "perennial gale of creative destruction" powered by innovation from "the new consumers' goods, the new methods of production or transportation, the new markets, [and] the new forms of industrial organization that capitalist enterprise creates."[5] Innovation has transformed the corporate landscape in only the past few decades by creating new industries such as internet search-driven advertising and social networking, and by adding value to more familiar industries such as personal computing and cellular communications.

Creative destruction poses a challenge to assess the implications of current market concentration for antitrust enforcement. Does a high market share justify concern about monopoly power, or does it reflect creative destruction from successful innovation? Will the gale of creative destruction make monopoly only a temporary phenomenon? If markets are undergoing rapid change, will a merger or acquisition raise future prices or harm innovation?

Creative destruction complicates predictions of market outcomes, but it does not make antitrust enforcement irrelevant or unnecessary. Long periods of calm can persist between the waves of creative destruction. Indeed, the average duration of firms in the Standard & Poor's (S&P) 500 Index, which includes many high-tech firms, has been increasing since 2000.[6] Although innovation can topple the corporate giants, there are also strong forces, such as high entry barriers and network effects, that can insulate firms in the high-technology economy from competition for decades.

**Many high-tech industries have high entry barriers**
High barriers protect many high-tech firms from new competition. Entry into these industries often requires large upfront R&D expenditures and production often has low marginal costs. Search engines have large fixed costs to develop algorithms, but the marginal cost of delivering another search result is close to zero. The combination of large up-front cost and low marginal cost (the cost of supplying another unit) makes new head-to-head competition a risky endeavor. Entrants into R&D-intensive industries face the risk that they will be unable to recover their R&D investments if established firms respond aggressively to their entry. Consequently, one or two firms often dominate the supply of many high-tech products and services. New entry, when it occurs, tends to be for products and services that do not replicate existing products and services.

Intellectual property protection is also a barrier to new competition in many high-tech markets. Many products and services in these industries are protected by patents, copyrights, or trade secrets. New competitors have to negotiate licenses to technologies controlled by incumbents or incur costs to invent workarounds that do not infringe the incumbents' technologies. Incumbents in some technological areas, such as computers, information technology, and biotech, control the rights to hundreds of patents that cover a product or technological application. These "patent thickets" create a formidable barrier to entry if new competitors have to negotiate with many different entities for licenses to these rights and do not have intellectual property of their own that they can trade. Furthermore, firms can allege trade secret theft to deter their employees from leaving to start potentially competitive new enterprises.

Traditional antitrust policy reflects the premise that society is better off when the price of a product or service is closer to its marginal production cost.[7] Antitrust policies, such as cartel and merger enforcement, often strive to prevent conduct that would increase prices or prevent prices from moving closer to marginal costs. That is consistent with the goal of achieving static economic efficiency for the use of existing resources. However, prices that are close to marginal costs often are not sustainable in markets with large R&D costs and consequently do not promote dynamic economic efficiency in these markets.

Sustainable prices must cover a firm's average production costs, but a firm's average cost typically exceeds its marginal cost when the firm incurs large fixed costs, such as R&D expenditures. Competition policies that force prices closer to marginal costs can extinguish incentives for R&D and other investments that power the technological engines of the modern economy. Many high-tech industries must have high profit margins to survive. While this requirement does not justify anticompetitive conduct, it implies that profit margins can be misleading indicators of market performance in these industries.

### Many high-tech industries have network effects that can entrench incumbents

Many firms that dominate sectors of the high-technology economy benefit from network effects that reinforce their market dominance.[8] Network effects are present when the value of a product to each consumer increases with the number of other consumers of the product. This can occur either directly, because consumers want to interact with other consumers of the same product or service, or indirectly, because more consumers attract more developers of complementary products that enhance consumer experiences.[9] Telephone and social networks exhibit direct network effects. Operating systems for computers and mobile devices exhibit indirect network effects because firms and consumers value the number of compatible applications, as well as their quality and diversity.

In markets with network effects, the success of a product or service depends on expectations about the number of firms and consumers that will adopt and support it in the future, in addition to its quality and price. Consequently, network effects can confer monopoly power on a firm that consumers expect will have large future sales. This phenomenon, in which a market tips to favor a monopoly supplier, can cause competition in markets with network effects to take the form of competition "for the market" rather than competition "in the market."

Firms can compete to win the market by aggressively cutting prices or by making exclusive deals with suppliers of complementary products that contribute to indirect network benefits.

Markets with network effects can exhibit excess inertia if firms and consumers fail to adopt a superior technology (i.e., a product or service) because the parties do not have confidence that others will do the same.[10] Technology adoption can also exhibit the opposite condition: insufficient friction.[11] That is, firms and consumers that adopt a new technology may ignore the effects of their adoption decisions on the installed base of an existing technology, thereby stranding users of an incompatible technology.

The best technology does not necessarily win a market with network effects. Excess inertia and insufficient friction can prevent the adoption of an efficient technology or cause the adoption of an inefficient technology if technologies are not compatible with each other.[12] Compatibility allows different technologies to benefit from network effects. If new and old technologies are compatible, the adoption of the new technology would not strand users of the old technology and concerns about the number of future adopters would not prevent firms and consumers from adopting the new technology. However, compatibility can incur costs and decrease innovation incentives by preventing the innovator from becoming an exclusive supplier.

Although network effects shape the evolution of many high-tech markets, they do not operate without limits. Network effects can attain a maximum value before the entire potential market is covered. These effects can turn negative and become congestion costs if the network includes members (such as spam marketers in a telephone network) that impose costs on the participants rather than conferring benefits. Professionals may place a high value on participating in a network such as LinkedIn, which has many subscribers from their professions and related professions, but they may not care much about the ability to interact with other members of the workforce. Decreasing returns to scale on the supply side can be a disadvantage for large firms relative to smaller upstarts and prevent markets with strong network effects from tipping to a monopoly supplier. Firms can offer consumers differentiated products, which can cause some consumers to prefer a product even if it has a much smaller network.

Network effects do not guarantee that a dominant firm will persist or that rivals cannot compete. Most music listeners transitioned from records to audiotapes, tapes to CDs, and CDs to streaming services, notwithstanding each previous technology's large installed base of consumers. Facebook replaced MySpace, Google leapfrogged Yahoo! and AltaVista, and Apple and Android smartphones supplanted the once-dominant Blackberry. Apple succeeded in the personal computing space despite the large installed base of applications that were compatible with Microsoft's operating system. Notably, these transitions embodied technological innovations or significant product differentiation; the successful entrants did not use lower prices or modest quality improvements to take market share from incumbents.

The distinctive features of markets with network effects motivate contrasting opinions about appropriate antitrust policies for markets with these characteristics.[13] One opinion is that antitrust enforcement should be tougher in the presence of network effects because they create high barriers to entry and entrench incumbents with market power. Network effects amplify the harm from exclusion because exclusionary practices can cause a market with network effects to tip to a monopoly. There is a contrary view that antitrust enforcement can lower consumer welfare by enabling firms to fragment the market, which would lower the maximum available benefit from network effects,[14] or that antitrust enforcement is unnecessary because the market will tip to a dominant supplier in any event.

Concerns about fragmentation in markets with network effects are not persuasive when the market has a dominant firm; it is conduct by a dominant firm that typically attracts antitrust scrutiny. Exclusionary conduct by a dominant firm is likely to harm consumers by preventing the entry of an efficient competitor or by suppressing its ability to compete. For this reason, antitrust enforcement should be more attentive to exclusionary conduct by dominant firms in markets with network effects and should apply a threshold level for anticompetitive exclusion that is less than substantial foreclosure.

The economic logic that underlies antitrust law for predatory conduct does not naturally extend to markets with network effects. Appropriate tests for predatory conduct are uniquely challenging for network industries because outcomes depend on consumer and firm expectations about future adoption decisions, and because firms often use low prices or other means to reward adoption and benefit from network effects. Consequently, network effects put antitrust authorities in the uncomfortable position of having to make guesstimates about expectations to justify enforcement decisions about predation allegations. Further research is needed to craft enforcement tools to address predatory conduct in network industries. For now, the best that can be said is that antitrust enforcers should address such allegations with caution.[15]

For the most part, antitrust enforcers have not employed a different rulebook for industries with strong network effects. As discussed in chapter 8, the court of appeals in *US v. Microsoft* introduced its opinion with the statement, "We decide this case against a backdrop of significant debate amongst academics and practitioners over the extent to which 'old economy' §2 monopolization doctrines should apply to firms competing in dynamic technological markets characterized by network effects."[16] However, with a few exceptions discussed in chapter 8, the court applied traditional antitrust policy regarding exclusionary practices by a firm with monopoly power and did not craft new policies to account for the fact that Microsoft operated in a dynamic technological market characterized by network effects.

### Many firms in the high-technology economy are platforms that serve populations with different price-cost margins

Unlike a traditional firm that provides a product or service to consumers, a platform provides products or services to two or more sides populated by different agents. Platforms are not new. Newspapers, over-the-air television networks, auction houses, videogame consoles, and real estate brokers are all examples of platforms. Platforms have gained prominence recently as a business model for the high-technology economy. Four of the five publicly traded US firms that had the highest market values on December 31, 2018 are technology platforms: Alphabet (Google), Amazon, Apple, and Microsoft, in that order.[17]

A platform is a more general case of a "two-sided market," a term that Jean-Charles Rochet and Jean Tirole introduced in 2002.[18] The term is a misnomer in some respects. Every market has at least two sides: It takes a seller and a buyer to consummate a transaction. Furthermore, "two-sided" markets often have more than two sides: Google's internet search engine provides services to consumers looking for information, advertisers that want to connect with consumers, and website publishers that want to showcase their products and possibly benefit from advertising revenues; Apple and Microsoft connect computer or smartphone users to application developers and device manufacturers.

Platforms differ from conventional markets in a number of respects, although the differences are often a matter of degree rather than a clear demarcation. In a platform, agents on at least one side of the platform have benefits that depend on the number of agents on the other side of the platform: these are "cross-platform" network effects. Facebook's advertisers value the number of participants on its social network. Restaurants that participate on OpenTable, a website that allows diners to make reservations, value the number of subscribers to OpenTable, and subscribers value the number of restaurants that use the service.

Another distinguishing characteristic is that a platform manager actively intervenes to determine prices and terms of service to incentivize participation on the various sides of the platform.[19] That may imply prices that are far above marginal cost for one side and zero or even negative prices (rewards or other inducements) for another side. In a conventional market, firms on one side of the market benefit from consumers on the other side, and vice-versa, but they make their pricing and participation decisions without active intervention from a separate party.[20]

Although several of the most prominent firms in the high-technology economy are notable examples of platforms with strong cross-platform network effects and active management, many conventional firms also display platform characteristics. For example, we don't typically characterize brick-and-mortar retail stores as platforms. Nonetheless, a shopping mall is a candidate for designation as a platform because there are cross-platform network effects (i.e., shoppers care about the number, quality, and variety of stores in the mall, and stores care about the number of shoppers and their shopping lists), and the mall manager may subsidize one or more anchor tenants in order to attract other stores.

Entry barriers can be very high for platform markets. Cross-platform network effects imply that a successful entrant must attract participants from at least one side of the platform, and often from multiple sides. A new operating system will not interest computer users if it does not have an attractive suite of applications, as IBM discovered when it attempted to compete head-to-head against Microsoft in the market for personal computer (PC) operating systems. Microsoft suffered a similar fate with its unsuccessful entry into the smartphone market.

Platforms that serve at least one side with zero or very low prices present an additional challenge for new competitors. A new search engine cannot compete by offering a lower price for search (short of paying consumers to join its platform), although it might offer a lower quality-adjusted price, such as by taking better care of user data. Differential markups of prices on different sides of a platform upend the usual signals to identify anticompetitive conduct. Search engines do not engage in predatory pricing merely because they allow consumers to search without charge and do not abuse market power merely because they charge advertisers a price that greatly exceeds the marginal cost of placing ads on the search engine results page.

Platforms can amplify the effects of exclusionary conduct. For example, suppose that a platform serves two sides, "A" and "B," and there is a single incumbent and a potential entrant. Suppose that the incumbent negotiates exclusive contracts with agents that participate on the A side. The contracts eliminate participation on the A side of the potential entrant's platform and therefore make it much more difficult for the entrant to compete on that side. Moreover, given cross-platform network effects, the contracts make the potential entrant's platform less attractive to participants on the B side and therefore make it more difficult for the entrant to compete on that side of the platform.

As discussed in the context of network effects, the consumer harm from exclusionary contracts depends on whether they are imposed by a dominant firm or a smaller rival. Exclusive dealing on either side of a platform can allow a dominant firm to prevent competition, but exclusive dealing can allow a rival to gain a foothold to challenge a dominant incumbent. DIRECTV's exclusive contracts with certain content providers (e.g., the NFL ticket) likely enhanced competition by making the satellite TV service an attractive alternative to cable.

In a similar fashion, a broad prohibition on exclusive arrangements can promote platform dominance. A requirement that products must interoperate with each other is similar in some respects to a prohibition on arrangements in which products are exclusive to a subset of customers. Robin Lee finds that compatibility across video games would allow the dominant video game platform to benefit from games developed for other video platforms and would reinforce its dominance. In his analysis, consumers who had previously multihomed by purchasing multiple game consoles, each of which had games that operate exclusively with a single console, would instead have purchased the dominant platform if all games and consoles were compatible.[21]

Exclusive arrangements can have contrasting competitive effects, which depend on the market power of the platform that imposes an exclusive arrangement and on the technical features of the affected markets. Broadly speaking, exclusionary practices by a dominant firm in a platform market have the potential to cause significant harm to competition and consumer welfare, but some forms of exclusionary contracts can promote competition from a smaller rival. Furthermore, a ban on exclusionary contracting does not necessarily promote competition or consumer welfare in platform markets.

Platform markets have characteristics that warrant careful attention for antitrust analysis, but do not generally require a different approach. The complaints filed in US v. Microsoft and in the European investigations did not identify PC operating systems as platforms, perhaps because the term was not yet established in the antitrust vernacular. Nonetheless, the complaints recognized that applications were a barrier to entry that protected Microsoft's monopoly. It is unlikely that more specific attention to platform economics in these cases would have supported different enforcement outcomes.

In contrast, the Supreme Court opinion in Ohio v. American Express faulted plaintiffs for failing to take full account of the two-sided nature of the market for credit card transactions.[22] Payment networks are clearly two-sided markets because transactions require a consumer that belongs to the network and a merchant that accepts payment from the network. The Supreme Court's opinion raised concerns that it heralded a fundamental shift in antitrust policy for platform markets, with onerous implications for antitrust plaintiffs.[23] Many firms can fit the description of a two-sided platform, and the complexity of a two-sided analysis of conduct can be a practical obstacle for antitrust enforcement. Does this opinion mean that every antitrust case that plausibly involves a platform will require a two-sided analysis?

The American Express case addressed the competitive effects of rules that prohibit merchants from steering customers to use a different credit card if the customer presents an American Express card for payment. The rules suppress competition at the point of sale by discouraging rival credit card networks from offering merchants better terms, such as lower transaction fees, to increase transactions on their networks and by preventing them from offering other inducements in return for promoting the use of their cards. The rules also encourage the issuers of credit cards to compete by offering their cardholders attractive awards to motivate the use of their cards. Issuers have incentives to reward cardholders for the use of their cards if they generate large merchant fees.[24]

The relevant question for antitrust analysis is not whether a market must be defined as one-sided or two-sided, but rather how best to measure the effects of conduct on competition and output. In some situations, a two-sided analysis is useful to address the trade-offs between competition that can occur on the different sides of a platform, but it is not necessary to address these tradeoffs.[25] Attention to competitive effects on one side of the platform can be appropriate, provided that the analysis fully accounts for relevant interactions with the other side, including cross-platform network effects to the extent that these interactions are significant. In the American Express case, the central issue is not whether the analysis must be two-sided, but instead whether a two-sided or one-sided analysis is more informative about the interactions between merchant rules and incentives for consumers to transact over payment networks.

Another example of an antitrust case involving a two-sided market is the Google Shopping case discussed in chapter 9. The Federal Trade Commission (FTC) in the US and the European Commission (EC) investigated Google's practices regarding its search results for comparison shopping services. The agencies were aware that Google operated a two-sided platform, but neither agency defined a two-sided price. The FTC declined to challenge allegations that Google was biasing its search results to disadvantage competition. The EC held that Google's search practices were anticompetitive. The EC considered the interaction between search results and the market for search-related advertising, although that interaction did not play a decisive role in its conclusions.

**Information is critical for the high-technology economy**
The high-technology economy is both a consumer and a producer of information. Society benefits when information is shared, but sharing reduces information's private value and the incentive to create knowledge. Unlike most conventional goods, where use by one entity denies use by others, information is a "nonrival" good that can be consumed by many. If someone eats an orange, that orange is not available to anyone else. But if someone develops an idea to create a better mousetrap, others can use that idea too. The nonrival characteristic of information creates a policy dilemma. Information is most valuable when it is freely available, but restrictions on copying are often necessary to allow firms to profit from creating the information in the first place.

Antitrust laws interact with intellectual property laws to provide incentives for innovation in the economy. Both are imperfect, and they are in tension with each other. Antitrust policies foster competition and oppose exclusionary conduct, while intellectual property laws confer rights that allow owners to exclude competition. It is often said that the two legal regimes are complementary because intellectual property provides incentives for innovation, which benefits consumers. While this notion is correct in some respects, intellectual property rights can also allow their owners to elevate prices and impose costs on innovators that build on existing knowledge to create new and improved products or services.

The design of optimal intellectual property protection is outside the scope of this book, but antitrust enforcement can affect the exercise of intellectual property rights. Intellectual property protection does not confer an exemption from antitrust enforcement for conduct that excludes competition. Antitrust agencies sometimes condition mergers in high-tech industries on agreements to license intellectual property rights. The agencies also can pursue remedies for monopolization that require a dominant firm to offer licenses to its intellectual property, including licenses required for firms to supply interoperable products.

High-tech firms such as Google and Facebook benefit from very large data troves that allow them to better target potential consumers for paying advertisers. Data, which some call the raw material of the digital economy,[26] is a nonrival good. If not shared, data can be a barrier to entry because it is costly to duplicate. Competition policy can address data as a barrier to entry, such as by requiring incumbents to make their data available to rivals or by giving individuals ownership rights to their data, notwithstanding protection of data as intellectual property.

Data has yet to raise unique antitrust issues. Although some very large data sets are important sources of value and potential barriers to competition,[27] relatively small data sets also can raise antitrust concerns, and large data sets need not foreclose competition.[28] The US Department of Justice (DOJ) required Thomson to sell copies of three financial data sets, license related intellectual property, and provide access to personnel and product support as preconditions to allow the company to merge with Reuters.[29] The DOJ concluded that the companies were two of only a few firms that supplied financial fundamentals data, earnings estimates data, and aftermarket research reports and alleged that without the divestiture the merger likely would have led to higher prices and reduced innovation for these products.[30]

The FTC raised competition concerns related to media audience data in its analysis of the merger of Nielsen Holdings and Arbitron. The EC alleged that the two companies were uniquely positioned to develop national syndicated cross-platform audience measurement services (i.e., aggregated measures of audience sizes for different media platforms) and required Arbitron to divest its cross-platform audience measurement business, including data from its representative panel, as a condition to approve the merger.[31]

The Thomson-Reuter and Nielson-Arbitron cases did not raise novel antitrust issues because the data sets were tightly connected to relevant products for which the agencies applied conventional analytical tools to analyze anticompetitive effects. The EC explored more general "big data" issues in its investigation of the acquisition of the professional social networking website LinkedIn by Microsoft. The EC considered whether the combination of large data sets controlled by Microsoft and LinkedIn would allow the merged company to increase its market power in a relevant market or increase barriers to entry. The EC found no cause for these more general concerns; nonetheless, the EC accepted commitments from Microsoft to facilitate interoperability of LinkedIn with other professional networks.[32]

### Complements, interoperability, and standards require different approaches to evaluate competition

Many high-tech industries support an ecosystem of firms that supply complementary products.[33] Applications run on computer and smart-phone operating systems. Specialized integrated circuits provide functionality to support microprocessors. Competition among suppliers of products or services that are complements differs from competition among suppliers of products or services that are substitutes for consumers or businesses.

When products A and B are substitutes, a lower price for A increases the demand for A and reduces the demand for B. If a firm only sells product A (or B) and lowers its price, it does not account for the reduction in the demand for a substitute product B (or A, conversely) and it is often delighted to benefit at the expense of its rival. A firm that sells both products would internalize this demand substitution and choose higher prices for A and B than firms would choose if they are sold separately. This is why mergers and cooperative price-setting raise flags for antitrust enforcers when the firms sell substitute products.

Suppose instead that products A and B are complements, such as operating systems and applications. Firms or consumers tend to use both products if they are complements. A lower price for product A increases the demand for product B by making the combination of A and B less expensive. A firm that sells only product A (or B) would not account for the positive effect of a lower price on the demand for B (or A, conversely). A firm that sells both products would internalize this effect and may choose lower prices for A and B than the sum of the prices for A and B that firms would choose if they are sold separately.

Competition lowers prices for products that are substitutes but independent price-setting can raise prices for products that are complements. The latter is the "Cournot complements effect," named after the brilliant nineteenth-century French economist Augustine Cournot.[34] The Cournot complements effect also applies to innovation incentives. A firm that supplies only product A would not account for the positive effect of an improvement on the demand for complementary product B, while an integrated supplier of A and B would internalize this cross-product effect.[35]

In many high-tech industries, the benefits from competition among firms that sell complementary products require that the products be capable of interoperating with each other. Interoperability in turn requires technical standards, which can come from coordination by formal standard-setting committees or from market forces. Committees created the different generations of cellular communication standards. Market forces supported the x86 architecture standard for PC microprocessors developed by the Intel Corporation.

Interoperability and the need for standards raise numerous potential antitrust concerns. Dominant firms may purposely obstruct interoperability to deter competitors. IBM was accused of strategically manipulating interoperability standards for connecting peripheral devices to its mainframes. Microsoft was accused of altering the Java programming language to prevent it from being able to support platform-independent applications. Interoperability concerns are not limited to the computer, internet, and information technology industries. Drug companies have been accused of patenting minor changes to drugs that are close to losing patent protection in order to prevent generic competition. The equivalence between generic drugs and their branded counterparts is a form of interoperability.

Formal standard-setting often can raise antitrust concerns, if only because it involves cooperation by actual or potential competitors. Established firms may agree to support standards that prevent competition from new technologies, and dominant firms may attempt to stack votes in standard-setting committees to adopt a favored standard. Other concerns relate to interactions between standards and intellectual property, because the adoption of a standard can cause patents that cover standardized technologies to have considerable market power. Formal standard-setting organizations and their participants can contribute to monopolization by failing to take actions to limit the power of these so-called standard-essential patents. The opposite is also a concern. Formal standard-setting organizations and their participants can be criticized for exercising buyer market power ("monopsony") by suppressing the ability of patent owners to achieve a financial return on their intellectual property.

Interoperability allows competing suppliers to benefit from industrywide network effects, and antitrust agencies have used their leverage to promote desired industry compatibility. Chapter 8 discusses interoperability conditions that were included in commitments that settled antitrust allegations against Microsoft. The Antitrust Division of the DOJ and the EC conditioned their decision not to challenge Cisco's acquisition of Tandberg on the parties' commitment to support an open industry standard.[36] The Cisco-Tandberg merger combined the two leading suppliers of "telepresence," a high-definition type of videoconferencing. Cisco agreed to divest ownership of its TelePresence Interoperability Protocol (TIP) to an independent industry body and authorized the industry body to license the rights necessary to implement TIP to any interested party, royalty-free.[37]

The FTC filed a complaint against Intel alleging that, among other conduct, the company stifled competition by impeding interoperability between Intel's newest microprocessors and competitors' graphics processing units (GPUs).[38] The FTC settled the case by issuing an order that, among other conditions, prevented Intel from making any engineering or design changes that would degrade the performance of a competing GPU without providing an actual technical benefit for Intel.[39]

A requirement to support an open standard or otherwise support interoperability often has economic justification. Firms with a large installed base of consumers can be biased against interoperability because they enjoy disproportionately large benefits from network effects, which bestow on them a strategic advantage vis-à-vis their rivals. However, an obligation to support compatibility with an open standard is not the best course for every circumstance. The chosen standard may not be the best technology, and compatibility can diminish competition that would have occurred when firms with incompatible products struggle to win market acceptance. Furthermore, compatibility can lower firms' innovation incentives because a share of the benefits from innovating would accrue to rivals that supply compatible products.

**Many high-tech firms have a low cost to integrate into new
markets and exclude competition**
Many firms in the high-technology economy face a relatively low bar
to expand their portfolio of products and services because they have
capabilities that apply to numerous products and services (called "economies of scope") and that allow them to produce and distribute new
products and services at low incremental costs.[40]

Suppose that General Motors (GM) wants to get into the business of
selling tractors. GM would have to invest in facilities to manufacture
the tractors and would need a distribution network to sell them. Tractors that fail to sell would be a costly write-off on GM's balance sheet.
Contrast GM's entry calculation with Microsoft's decision to supply
web browsing functionality along with its Windows 95 operating system
and subsequently integrate web browser code in Windows 98, discussed
in chapter 8. Although Microsoft invested heavily to develop its Internet
Explorer (IE) web browser, it was relatively easy for the company to
deliver IE to consumers even before downloading became a popular
distribution method. The cost of including IE on a compact disk was
negligible, and Microsoft did not incur additional costs if consumers
chose not to use the browser. Another example is Google's development
of a comparison shopping service. A shopping service can utilize many
of the same algorithms and hardware as a general internet search engine.
The service incurs negligible incremental costs once developed and is
distributed along with other Google search results.

Consumers may benefit when a dominant firm expands into new or
related markets, while rivals may condemn the expansion as anticompetitive leveraging of monopoly power. Distinguishing procompetitive
expansion from anticompetitive leveraging is a critical challenge for
antitrust enforcers. Chapter 9 describes the different responses by the
US FTC and the EC to Google's expansion into the display of comparison shopping services in its search results. The comparison shopping
example shows that enforcers have not reached agreement on how to
deal with this complex issue.

The high-technology economy raises other concerns, such as the
protection of privacy, that this book does not address because they do
not fall within the conventional scope of competition policy. Nonetheless, it would be wrong to conclude that these concerns have no relevance for antitrust enforcement. Firms compete in dimensions other
than price, and the protection of privacy is a non-price dimension that

is similar to product quality. For example, Facebook can increase its dominance of social networking by promising to keep consumer data private while subsequently breaking that promise to generate additional advertising revenue.[41] Such a promise could have antitrust implications by allowing Facebook to increase or maintain its dominance of social networking for reasons unrelated to competition on the merits.

Furthermore, there is scope for regulations, such as rights to personal data, that could facilitate competition by making it easier for consumers to switch to alternative suppliers, much like number portability facilitates switching between cellular networks. A more controversial requirement would compel companies to share their data with others at regulated terms. The European Union's recent General Data Protection Regulation obligates companies that operate in Europe to protect personal data and gives consumers greater control over the use of their data. These regulations should be designed with care because they can have unintended adverse competitive consequences by imposing compliance costs that are difficult for smaller rivals to bear and by creating additional barriers for new entrants into data-intensive activities if consumers do not choose to share their data.

### 3   Does the High-Technology Economy Need More Aggressive Antitrust Enforcement?

Innovations over the past few decades have delivered astounding improvements in computing power, communications, and connectivity. Yet a number of indicators suggest that entrepreneurial vigor is lacking in the high-technology economy and the benefits from technological progress have not been shared broadly with the American workforce. Several firms in the high-technology economy have earned high returns on invested capital with relatively modest increases in investment.[42] Increases in profitability have coincided with a decrease in the share of output claimed by workers, which aggravates inequality.[43]

Should antitrust enforcement take more active measures to address the increase in market power in the US economy? There is a flood of opinion statements calling for more aggressive antitrust enforcement, including breakups of tech titans, and for changes in antitrust standards to support these actions.[44] Aspiring politicians call for antitrust enforcement with a passion that we have not seen for over a century.[45] A focus on innovation could energize antitrust enforcement, but it should not abandon the fundamental tenet that big is not necessarily bad. Consumers benefit when firms develop new products or disrupt markets by employing more efficient methods of production and distribution.

Nonetheless, consumers can benefit from a change in the direction of antitrust enforcement to focus on innovation. Such changes include a lower standard for anticompetitive exclusion than substantial foreclosure in markets such as computer operating systems where network effects reinforce dominance. Antitrust policy also should be more vigilant to prevent acquisitions by dominant firms that may challenge their dominance.

Merger enforcement is the policy lever that antitrust authorities use the most to influence industry evolution, and there is evidence the authorities have tolerated too many mergers in recent years.[46] A theme in this book is that antitrust enforcement should focus on conduct that eliminates potential competition, particularly acquisitions that silence potential innovators. There is a decline in new business formation. Since 2000, initial public offerings (IPOs) have recently declined by more than one-half compared to the period 1980–2000.[47] The number of publicly traded firms has declined along with the decline in the number of IPOs. The Wilshire 5000 stock market index—a list of the 5,000 largest publicly traded firms—contained only 3,816 stocks in 2017.[48] As with other economic indicators, there are many explanations for this trend, among which are increasing financial regulation of public companies and the availability of private capital. But one explanation is merger policy that has allowed dominant firms to acquire hundreds of start-ups that might have floated public offerings as new companies.

Venture capitalists describe a "kill-zone" that surrounds the major tech companies. They allege that the tech giants acquire start-ups before they can become a competitive threat or they copy and trample firms that attempt to compete, and consequently they are reluctant to fund start-ups that intend to compete in applications that overlap with the core competencies of the major tech companies. After Snap Inc. rebuffed Facebook's attempt to acquire its multimedia messaging app, Snapchat, Facebook cloned many of Snapchat's features.[49] Amazon introduced its own home intercom system and video-conferencing tool after it invested in the start-up Nucleus, which sells a similar product.[50]

The tech companies can spot potential rivals in their infancy and have sophisticated tools to identify promising technologies.[51] Google and its parent company, Alphabet, made more than 200 acquisitions between 2001 and 2017.[52] They include the Android operating system;

YouTube; Motorola Mobility for smartphones and related intellectual property;[53] Zagat for restaurant reviews; the Quickoffice productivity suite; Waze for navigation; the mobile payment company Zetawire; the Picasa image organizer; Nest for home internet-connected devices; ITA Software, which provided the flight information used by many airlines and travel sites; and several artificial intelligence (AI) firms, including Deep Mind, Dark Blue Labs, Halli Labs, and Kaggle.

Google is the dominant supplier of search-based advertising. Acquisitions have helped Google to acquire a leading position in technologies that service advertising for internet and mobile networks. They include: the ad serving intermediary DoubleClick, which Google acquired in 2007; AdMob, the leading ad network for advertising on mobile platforms, acquired in 2010; Invite Media, which manages the purchasing of advertising inventory, also acquired in 2010; and the ad inventory manager AdMeld, acquired in 2011.[54]

Facebook made more than sixty acquisitions between 2005 and 2017. They include the social networking site ConnectU; the social network aggregator FriendFeed; Beluga for group messaging; Oculus VR for virtual reality; the ad server LiveRail; the photo-sharing site Instagram; the mobile instant messaging service WhatsApp; and the social polling app tbh, which allows users to create surveys and send them anonymously to their friends. WhatsApp had more than 400 million active users when Facebook bought the company for $19 billion in 2014. When Facebook acquired tbh (as in "to be honest") in 2017, the app had surpassed 5 million downloads and had more than 2.5 million active users, even though it had been available for only a few months. Despite its initial popularity, Facebook shut down tbh soon after the purchase, which Facebook explained was due to low usage of the app.[55]

Other tech titans strengthened control of their markets and ventured into new markets with numerous acquisitions. Apple made more than ninety acquisitions between the date the company went public and 2017. Recent acquisitions include a cloud service firm; Shazzam for music and image recognition; and a number of start-ups that specialize in AI, including RealFace for face recognition and Lattice for data mining. For its part, Microsoft has made hundreds of acquisitions, including the videoconferencing service Skype; the telecommunications and consumer products company Nokia; and the open-source software development platform GitHub.

Amazon is also an active player in the acquisition game. The long list of Amazon acquisitions includes online bookstores in Germany and the UK; the internet movie database IMDb; the online music retailer

CDNow; the online software retailer Egghead Software; the online shoe and apparel retailer Zappos; Quidsi, which operates websites for consumables such as diapers.com and soap.com;[56] and the grocery chain Whole Foods.

The semiconductor chip giant Intel made fewer acquisitions compared to the newer tech titans for most of its history, but it recently jumped on the acquisition bandwagon. The growth of AI spawned a new generation of chip-makers with designs that emphasize application-focused performance. Intel acquired Nervana Systems, a new entrant into AI-specific chips, in 2016. Intel also acquired Mobileye, a developer of advanced driver-assistance systems, presumably as a complement to a move into AI.[57]

Several of these acquisitions cemented the dominance of these giants and should have been reviewed with greater attention to their likely future effects. Google's acquisition of ITA Software allowed the company to improve its Google Flights service, but the acquisition also may have hobbled competition from other travel sites such as Expedia or Kayak. Furthermore, by making ITA Software captive to Google and its Alphabet parent, the acquisition may have lessened incentives to improve the product that ITA would have had as an independent company that served the entire online travel industry.

The US and European authorities reviewed, but did not challenge, Facebook's acquisition of WhatsApp and Instagram. The EC noted that WhatsApp and Facebook were but two of many messaging services and that WhatsApp did not compete with Facebook for online advertising.[58] Both agencies should have paid greater attention to the possibility that WhatsApp could have become a rival social network, much as the multipurpose messaging service WeChat has done in China (albeit censored by the authorities). Indeed, the $19 billion that Facebook paid for the app, despite little usage at the time in the US, should have been an indicator of its potential as an industry disruptor.

Instagram had thirteen employees and no advertising revenue when Facebook acquired the company in 2012 for approximately $1 billion in cash and stock. The UK Office of Fair Trading reviewed the acquisition and concluded that Instagram was one of many competitors in the photo app space and was not uniquely placed to compete against Facebook, either as a potential social network or as a provider of advertising space.[59] That prediction failed to appreciate Instagram's potential both as a type of social network and as a platform for advertising. By 2018, Instagram was generating approximately $7 billion in advertising revenue and had become a major source of revenue growth for

Facebook.[60] As in Facebook's acquisition of WhatsApp, antitrust enforcers again overlooked the potential of an emerging service to disrupt the acquiring firm's market.

Google's acquisition of the mobile advertising firm AdMob in 2010 is another illustrative case study of the failure of antitrust enforcers to prevent acquisitions that ultimately reinforced dominance in the digital economy. Mobile application developers and publishers rely on mobile ad networks to sell advertising space that they cannot sell effectively on their own. AdMob and Google were the leading mobile advertising networks when the FTC opened its investigation of the proposed acquisition in late 2009. The FTC was poised to block the transaction until it learned that Apple was about to enter the market with its own mobile advertising network, iAd. The FTC concluded that as the owner of the dominant mobile platform, Apple was uniquely positioned to compete with Google and AdMob. As a result, the FTC reversed course and allowed Google to acquire AdMob.[61]

Perhaps the FTC should have followed its prior instincts. According to eMarketer, in 2014, Google's share of US mobile ad revenue was about 37 percent and Apple's share was less than 3 percent.[62] In 2019, the consulting service Mobbo reported that Google's AdMob serves about 83 percent of ads on Android phones and 78 percent on Apple iOS phones.[63] These numbers are not sufficient to conclude that Google's acquisition of AdMob harmed competition that otherwise would have occurred for mobile advertising. That would require an in-depth analysis of the state of competition in the industry, not just a snapshot of market shares. Instead, they show that predicting future competition is a risky business, and antitrust enforcers should be aware of the risk of making errors of allowing harmful acquisitions to occur, as well as the risk of overenforcement.

The US antitrust agencies blocked a few acquisitions in high-tech sectors of the economy and approved others with conditions, but most acquisitions of small companies went unchallenged. It is likely that the agencies missed some important acquisitions, which are hard to unwind once integrated into a corporate structure. The Hart-Scott-Rodino (HSR) Act requires parties to notify the antitrust agencies about proposed mergers if revenues exceed set thresholds. That has no bite for acquisitions of start-ups with little or no revenues. Furthermore, many acquisitions by major tech platforms are not companies, but rather are people (called "acquihires") with skills that can be critical to new applications.

Acquisitions, even small ones, can eliminate a significant source of competition and innovation. An empirical study of the effects of pharmaceutical acquisitions (aptly titled "Killer Acquisitions") found that companies were more likely to terminate drug research projects that they acquired from other companies than projects that they originated themselves when the acquired projects were similar to projects in the acquiring firm's R&D pipeline.[64]

Nonetheless, it would be unwise to categorically prohibit dominant firms from acquiring start-ups that operate in the same or related businesses. Preventing the acquisition of a potential competitor runs the risk that the acquired firm may never have evolved as a significant competitor or may not have made as significant a contribution to the economy without the acquisition partner. Dominant firms have assets and capabilities that can energize new technologies, and the promise of a lucrative buyout is a major stimulus for start-up innovation. Although some entrepreneurs have the ability to create independent public companies, many others were formed with a narrow goal to develop a specific capability that would attract the interest of an incumbent firm. For these start-ups, becoming stand-alone companies is not necessarily the best way to ensure commercialization of their innovative technologies.

Whether consumers benefit from these acquisitions or whether they merely buttress the market power of existing dominant firms is a critical question for antitrust enforcement. Instagram and YouTube benefited from the market power of the platforms that acquired them, but they could have offered advertisers and internet users additional alternatives if they had remained independent companies or become affiliated with other internet entities. Chapters 5 and 7 explore innovation concerns related to mergers and acquisitions of potential competitors in more detail.

## 4   Issues for Applying Antitrust Law to Innovation

The antitrust laws do not need to be changed to address innovation.[65] The Antitrust Modernization Commission, established by Congress to consider whether the antitrust laws are adequate to address consumer welfare in the global high-technology economy,[66] concluded, "There is no need to revise the antitrust laws to apply different rules to industries in which innovation, intellectual property, and technological change are central features."[67]

Although the antitrust statutes are broadly flexible, courts have applied them in ways that raise barriers to the enforcement of dynamic competition. The tools that courts have developed to evaluate antitrust allegations, such as market definition, focus on price effects in existing markets. This policy evolution brought commendable analytical rigor to antitrust enforcement. Yet the price-centric analytical focus of modern antitrust enforcement also erects hurdles to evaluate allegations of harm to innovation. No court in the US or Europe has litigated to a final verdict an antitrust case that deals solely with innovation. This section addresses some of the obstacles to litigation of innovation effects. The focus here is on mergers, although there are similar obstacles for enforcement of single-firm conduct that affects innovation.

### Antitrust market definition

Section 7 of the Clayton Act prohibits mergers if "in any line of commerce or in any activity affecting commerce in any section of the country, the effect of such acquisition may be substantially to lessen competition, or to tend to create a monopoly."[68] In 1956, the US Supreme Court affirmed that the phrase "in any line of commerce or in any activity affecting commerce in any section of the country" means that competitive effects must occur in a "relevant market." In *United States v. E. I. DuPont de Nemours*, the Court said:[69]

Determination of the relevant market is a necessary predicate to a finding of a violation of the Clayton Act because the threatened monopoly must be one which will substantially lessen competition "within the area of effective competition." Substantiality can be determined only in terms of the market affected.

A rigid interpretation of the relevant market requirement for the Clayton Act is fatal for the evaluation of mergers that may affect innovation or future competition in markets that do not presently exist. R&D is not bought and sold in a market, apart from contracted R&D, but that does not mean that mergers cannot harm innovation by reducing incentives to invest in R&D for new or improved products. A merger of firms that are working to develop similar new products could have adverse effects on future price competition if both companies would have been successful innovators. Courts may not consider such future price effects because they occur in markets whose boundaries are inherently uncertain. Yet, a refusal to consider the possibility of these future price effects is tantamount to ignoring them entirely.

Jurisprudence on mergers and future competition is limited, but the available precedent is unhelpful for the analysis of innovation competition. In *SCM Corp. v. Xerox Corp.*, a district court heard a complaint that Xerox engaged in various practices that excluded SCM from markets for plain- and coated-paper copiers, including acquisitions of patents.[70]

While noting that patent acquisitions are not exempt from § 7 of the Clayton Act, the judge concluded that "liability for retrospective money damages cannot be predicated under § 7 upon a patent acquisition made prior to the existence of a relevant product market.... Indeed, there is considerable doubt whether liability can be grounded under § 7 for the mere acquisition of any asset prior to the existence of a relevant product market."[71]

The court of appeals affirmed the ruling and held that, "*The existing market* provides the framework in which the probability and extent of an adverse impact upon competition may be measured."[72] More recently, a district court rejected a market for R&D because the complaint did not identify one or more product markets consisting of reasonably interchangeable goods.[73] The court of appeals affirmed that ruling.[74]

Fortunately, the antitrust enforcement agencies have not followed the opinions expressed in *SCM v. Xerox* and its progeny. Chapter 7 describes several proposed mergers for which the agencies have alleged anticompetitive effects in markets that do not presently exist, or for which the contours are not sufficiently established to clearly define a relevant market. But it remains to be seen whether these challenges would survive if litigated in the courts.

A related issue is whether the pace of innovation in high-tech markets renders market boundaries so fluid that conventional antitrust analysis, which focuses on competitive effects in relevant markets, becomes too speculative for sound enforcement decisions. Although the potential for disruptive change may ease concerns about anticompetitive conduct under some circumstances, the possibility that markets may differ in the future does not prevent anticompetitive harm in the present or prevent those harms from having lasting effects.

Section 7 of the Clayton Act is intended to curtail the anticompetitive consequences of mergers in their incipiency. It addresses anticompetitive effects that are likely to occur in the future but cannot be predicted with certainty. Limiting application of the act to competitive effects that occur in existing markets is unnecessarily restrictive and contrary to the purposes of the act.

### Potential competition
Suppose that a drug manufacturer acquires a company that has a drug in development, which, if successful, would compete with a drug supplied by the acquiring firm. The acquisition would potentially eliminate an independent competitor in a market presently occupied by the acquirer. Courts have imposed high hurdles to establish antitrust harm

from potential competition when the competition is uncertain,[75] but this policy is not appropriate for mergers in which the potential competitor is a potential innovator. If the acquisition was not an important motivator for the innovation in the first place and has no efficiency benefits, it should be blocked if it would eliminate an important competitor in the event of successful innovation by the acquired entity, even if the probability of success is small.

I discuss these issues in more detail in chapter 5. In some cases, acquisition by an established firm is the most powerful incentive for innovation and the best way to commercialize a new product. Prohibiting acquisition by an established firm could curtail innovation by preventing the commercialization of the product. In other cases, acquisition of a potential innovator by an established firm should be blocked because it risks eliminating an important independent competitor with no offsetting benefits. A competition policy that balances the potential for acquisitions to motivate innovation against the harm from suppressing independent competition would allow antitrust enforcers and the courts to prevent acquisitions if there are other likely acquirers that are not actual or potential competitors.

**Efficiencies and appropriability**
Mergers and other arrangements can have benefits for innovation that differ from the types of benefits that antitrust authorities may recognize (albeit grudgingly[76]) and credit against possible anticompetitive effects. For example, antitrust authorities are more likely to value savings in variable production costs rather than savings in fixed costs because the former are more likely to be passed on to consumers in the form of lower prices. Antitrust authorities may dismiss R&D cost-savings because they are savings in fixed costs, and they may regard them as unacceptably speculative because any consumer benefits from the savings occur in the distant future—and maybe never. However, just as antitrust authorities should not dismiss innovation concerns merely because they cannot be proven with a high degree of certainty, they also should not apply an unreasonable standard of proof to acknowledge R&D cost-savings and related efficiencies, including benefits from conduct or transactions that allow an innovator to appropriate greater value from its R&D efforts.

It is possible, if not likely, that a merger or other arrangement can promote innovation while also raising prices. If courts follow their historical practices, they are unlikely to trade off R&D efficiency and innovation benefits against higher prices, even if the benefits could conceivably compensate for the consumer costs from higher prices.[77] Courts should reconsider this historical approach to provide a level playing field to evaluate innovation benefits from mergers and other conduct.

### Evidentiary requirements

Antitrust enforcement has moved beyond structural presumptions based on market shares and increasingly relies on quantitative analysis of competitive effects, particularly for mergers. The demand for empirical precision is commendable, but it is also an obstacle for plaintiffs that allege harm to innovation or for defendants that allege innovation benefits, because it is difficult to construct empirical tests that isolate the effects of a transaction on innovation.

Lack of precision should not prevent courts and antitrust authorities from considering innovation effects when there is a reasonable basis for an enforcement decision. Plaintiffs bear the burden to prove their allegations, and it is socially costly to hold parties liable for conduct that is unlikely to create harm. But it would be unwise to abandon merger enforcement for innovation merely because effects cannot be predicted with a high degree of certainty. Decisions not to challenge mergers that may harm innovation are also subject to error and can have high resulting social costs.

Absent analytical tools, corporate documents and industry testimony can sway enforcement decisions. In a merger of companies X and Y, that can include statements such as, "We [company X] need to keep innovating to keep up with Y." However, companies can control what their executives put in print, and industry testimony can be self-serving. A better alternative is a body of theory and empirical validation that courts can rely on to make predictions of alleged innovation effects.

Structural presumptions for innovation and future price effects should be justified by empirical research in related industries, which measures the effects of historical merger activity in an industry on innovation outcomes. The empirical studies in turn should be supported by sound theoretical analysis. This book is an effort to assemble a reliable reference for the theory and empirical research to justify antitrust policies that balance the risks of overenforcement and underenforcement for innovation and competition for new products and services.

# 3 Competition and Innovation Basics: Arrow versus Schumpeter

What we have got to accept is that [the large-scale establishment or unit of control] has come to be the most powerful engine of [economic] progress and in particular of the long-run expansion of total output.... In this respect, perfect competition is not only impossible but inferior, and has no title to being set up as a model of ideal efficiency. It is hence a mistake to base the theory of government regulation of industry on the principle that big business should be made to work as the respective industry would work in perfect competition.
—Joseph A. Schumpeter, *Capitalism, Socialism, and Democracy* (1942)

The incentive to invent is less under monopolistic than under competitive conditions, but even in the latter case it will be less than is socially desirable.
—Kenneth Arrow, "Economic Welfare and the Allocation of Resources for Invention" (1962)

## 1 Introduction

This chapter describes two basic economic propositions regarding the relationship between competition and innovation. Section 2 describes a view attributed to Joseph Schumpeter that imperfectly competitive markets promote innovation by allowing firms to appropriate more value from their innovations.[1] Section 3 describes a theory, developed by Kenneth Arrow two decades later,[2] that implies that a firm with monopoly power that is protected from competition has less incentive to innovate than a firm in a competitive industry. The monopolist's existing profits discourage research and development (R&D) if an innovation risks the loss of these profits. Firms in a competitive industry do not have a corresponding opportunity cost from innovation. Arrow's theory supports a statement by Sir John Hicks that "the best of all monopoly profits is a quiet life."[3] According to this theory, a firm that is shielded from competition has neither the necessity nor the incentive to work hard to create new and improved products. The loss of existing profits from an innovation is often called the "Arrow replacement effect." It is a fundamental force that appears repeatedly in this book.

Section 3 explores assumptions in the Arrow model that limit its generality. Arrow does not consider the implications of limited appropriability discussed in section 2. Furthermore, he does not consider incentives for the monopolist to innovate to protect its profits by excluding rivals. Moreover, he only examines cost-reducing innovations. Competition can have different implications for incentives to develop new products compared to incentives to lower the cost of producing existing products.

Neither Arrow nor Schumpeter addresses the dynamic interactions that exist between incentives to innovate, the production of innovations, and the state of competition in markets. Chapter 4 discusses these dynamic interactions, along with organizational theories that depart from simple profit maximization. I also note here that the effects of reduced competition on the incentive to innovate are not generally equivalent to the effects of a merger on innovation incentives. A merger places two formerly independent entities, along with their premerger R&D assets, under centralized control. Absent merger-specific economies or diseconomies of scale or scope in R&D, a merger does not change an industry's capacity to invest in R&D. Instead, a merger allows the merged parties to coordinate economic decisions, such as setting prices and investing in R&D. Chapter 5 explores the effects of mergers on innovation incentives in more detail, and chapter 6 examines the empirical evidence relating competition and mergers to innovation.

## 2   Joseph Schumpeter and the Consequences of Limited Appropriation

In the 1940s, Joseph Schumpeter challenged the prevailing view that competition is a superior form of market organization to create social value. In regard to dynamic efficiency and the creation of new products, Schumpeter wrote that, "Competition is not only impossible but inferior, and has no title to being set up as a model of ideal efficiency."[4]

Schumpeter did not speak directly to the effects of market structure on the ability to appropriate the value of innovations. He was more concerned with the failings of models of perfect competition to account for innovation and entrepreneurship and emphasized the power of "creative destruction" to invigorate economic progress, which can come from any source.[5]

Nonetheless, Schumpeter's name has become shorthand for the proposition that scale and market power enable a more stable and productive platform for R&D. In *Capitalism, Socialism and Democracy*, he concludes that:[6]

There are superior methods available to the monopolist which either are not available at all to a crowd of competitors or are not available to them so readily: for there are advantages which, though not strictly unattainable on the competitive level of enterprise, are as a matter of fact secured only on the monopoly level, for instance because monopolization may increase the sphere of influence of the better, and decrease the sphere of influence of the inferior, brains, or because the monopoly enjoys a disproportionately higher financial standing.

Schumpeter provided no formal economic model to describe his vision of monopoly power and innovation incentives, and his emphasis on the ability of large firms to attract capital is outdated, given developments in the availability of venture capital. Nonetheless, his criticism of perfect competition as the ideal engine of innovation is valid. Perfect competition is not viable in industries such as semiconductor fabrication, for which new facilities incur multibillion-dollar sunk costs, or in industries such as computer software or genomics, for which the marginal cost of technology licensing and distribution is only a small fraction of the R&D costs required to create the licensed products. Pricing these products at or close to their marginal production costs would not generate sufficient profits to justify the R&D expenses that brought them to market.

According to this Schumpeterian perspective, a reduction in rivalry can enhance appropriation in several ways. For example, a reduction in rivalry can allow an innovator to profit from a higher profit-maximizing price for the innovation. In that case, antitrust enforcers would have to weigh the adverse effects on prices from a reduction in rivalry against possible benefits for innovation. In this section, I address two other ways in which a reduction in rivalry can enhance appropriation: (1) A reduction in rivalry can increase an innovator's sales, which allows an innovator to appropriate more profit from the innovation; and (2) a reduction in rivalry can reduce technological spillovers, which allows competitors to imitate an innovation at the expense of the innovator.

To illustrate the effect of firm size on appropriation and innovation incentives, compare a highly stylized example of innovation by a firm in a duopoly to innovation by a monopoly. Suppose that there is a market in which price is fixed at $10 and total demand is fixed at 1 million units. There is an existing technology that enables production to take place at a constant marginal cost of $8 per unit, but a new technology can lower that cost to $6. Furthermore, suppose that the innovator cannot profitably license the new technology, perhaps because it would be difficult to monitor and prevent unauthorized copying.

Each firm has a profit margin of $2 without the innovation and $4 with the innovation. In the duopoly, each firm sells 500,000 units, and the innovation would add $1 million to each firm's profit. A monopolist sells 1 million units, so innovation by the monopolist would add $2 million to the firm's profits. The monopolist has a greater incentive to innovate because it can appropriate more of the innovation's value.

This simple example makes many assumptions. More generally, a successful innovator can profitably choose a lower price. A lower price would expand total sales and, in the duopoly case, allow a successful innovator to win sales from its rival. A monopolist also can benefit from greater total sales, but the monopolist has no rivals from which the firm can take sales. Furthermore, a new entrant or a small innovator might profit by licensing the innovation to other firms.

Partha Dasgupta and Joseph Stiglitz developed a formal model of cost-reducing innovations in an oligopoly,[7] and Xavier Vives developed the analysis further.[8] Both models show that there are circumstances in which firms in concentrated markets can have greater incentives to invest in cost-reducing innovations than do firms in less-concentrated markets. The benefit from a cost reduction is proportional to a firm's output, and the larger the firm's scale, the more it can benefit from cost-reducing innovation. Competition lowers prices and may increase total output, but it fragments output among firms that share the market with their rivals. By doing so, competition can reduce firm scale and lower the incentive for each firm to innovate.

This discussion of appropriation and R&D investment exposes another issue that is relevant to competition policy: the distinction between R&D investment and innovation. Greater R&D investment does not necessarily imply more innovation. Indeed, the correlation can be negative. For example, suppose that it costs $1 million to reduce the marginal production cost from $8 to $6. A monopolist incurs an R&D cost of $1 million for the cost reduction, while in this example, the industry with two firms incurs twice the R&D cost for the same improvement in the total cost of production, assuming no difference in industry demand. There is greater R&D expenditure in the duopoly market compared to the monopoly, but no greater benefit for innovation.

The possibility that rivals can gain valuable information from another firm's innovations (i.e., technological spillovers) can reduce the incentives of firms in competitive markets to invest in R&D. Patents, trade secrets and copyrights provide only limited protection from imitation by rivals. Although the threat of imitation can reduce a firm's incentive to invest in R&D, it does not necessarily reduce innovation. Spillovers extend the benefit of innovation by allowing innovation by one to be enjoyed, at least partially, by others. With spillovers, a competitive industry may not have adequate incentives to invest in R&D, but the innovation that occurs yields high returns because it benefits many producers.[9] Spillovers are not always beneficial in this respect because they can encourage firms to play a waiting game, in which each firm postpones R&D in the expectation that it can learn from successful innovation by its rival. In extreme cases, the waiting game can result in a losing scenario, in which firms postpone innovation indefinitely.[10]

If an industry is characterized by high technological spillovers, an efficient policy is to subsidize R&D to increase the incentives for investment while encouraging competition to keep prices low. Without subsidies for R&D, an increase in market concentration can be a second-best solution, by reducing the number of rivals that can benefit from spillovers and thereby increasing incentives for innovation. Antitrust authorities, who have no control over R&D subsidies, then would have to balance the increased incentive for innovation from greater market concentration against the risk of higher prices.

I turn next to a model of the effects of competition on innovation incentives developed by Kenneth Arrow two decades after Joseph Schumpeter's writings on this issue.

## 3   Arrow's Model: The "Replacement Effect"

Arrow examines a firm's incentive to invest in a cost-reducing innovation in two polar market structures: (1) a monopolist that is protected from competition and (2) a perfectly competitive market. Because Arrow's theory is fundamental and influential, I review his model in detail in this section. Arrow makes the following assumptions:

- In both market structures, there is an existing technology that allows the production of a product with constant marginal cost.
- The innovation is a new technology that allows production of the same product at a lower constant marginal cost.
- The innovation is protected by a patent that prevents imitation.
- The product is sold at a uniform price and the profit from the innovation does not depend on the identity of the innovator.

Arrow shows that the incentive to invest in the new technology is greater for a firm in the competitive market than it is for the monopolist. The monopolist has an existing flow of profit from the old technology, which the invention would replace. A firm in a perfectly competitive market has no existing profit flow that is at risk from an innovation.

The fact that a firm in a competitive market has more to gain from the new technology than a monopolist is obvious if the new technology eliminates competition from the old technology. Arrow calls this a "drastic" innovation.[11] The monopolist and a firm in a competitive market earn the same profit with the innovation, but the firm in the competitive market has nothing to lose by innovating.

Incremental (nondrastic) innovations take sales from—but do not eliminate—competition from existing products or from the use of existing technologies. Arrow shows that a firm in a competitive market also has a greater incentive to invest in R&D for an innovation that reduces the marginal cost of production but is not drastic. His insight is that monopoly prices reduce sales compared to a firm in a competitive market for any level of the firm's cost. Given reasonable assumptions about demand for the product, the lower sales by the monopolist reduce its incremental benefit from a cost reduction compared to the benefit for a firm in a competitive market.

Arrow concludes that the incentive to innovate is less than its social value in both the competitive and the monopoly regimes. This follows from his assumption that the innovator charges a single uniform price for the product, which implies that the monopoly profit from the innovation is less than its social value: Some consumers who would benefit by consuming the product are excluded because its price exceeds its marginal cost, while others purchase the product at a price that is less than its value. If the monopolist could capture all of the social benefit from an innovation, its incentive to innovate would equal the innovation's social value.[12]

Arrow's model applies to any firm that has existing profits and is protected from competition. Its significance cannot be denied, but it is also important to understand its assumptions and limitations. In addition to the assumption that the product is sold at a uniform price, they include the following:

• Arrow's monopolist is protected from competition in both production and R&D. Consequently, the incumbent monopolist has no need to engage in conduct to prevent potential competitors from innovating, including by being the first to invent and patent the new technology.
• The innovation in Arrow's model is a reduction in the marginal cost of production. His model does not address the more typical case of product innovation. Consequently, his analysis does not consider possible interactions between an innovation and the profitability of a firm's existing product portfolio, including the incentives of a firm to increase sales by improving a durable good.
• The innovator in Arrow's model benefits from perfect patent protection. He does not consider the relationship between competition and innovation incentives when imperfect protection limits the ability of the innovator to profit from invention.
• Arrow does not consider cumulative innovation, in which firms make R&D investments that benefit from other firms' prior innovations, nor does he consider how market structures change over time to reflect the success of past innovations.

Imperfect patent protection invokes issues related to appropriation discussed in the preceding section, as does his assumption that the product is sold at a single uniform price. I discuss cumulative innovation and dynamic models of innovation incentives in the next chapter. In this section, I address how Arrow's conclusions change if a monopolist has incentives to innovate in order to exclude rivals, and if innovations are new products rather than new production technologies.

### Arrow assumes a monopoly in both production and innovation

The monopolist in Arrow's monopoly regime faces no competition in either production or innovation. If the monopolist chooses not to invest in R&D, it retains a monopoly with the old technology. Eliminating the assumption that a monopolist is protected from innovation competition can reverse Arrow's conclusion that firms in competitive industries have greater incentives to invest in R&D.

A firm with monopoly power has an incentive to *preempt* competition that would upset its monopoly.[13] Andy Grove, former chief executive officer (CEO) of Intel Corporation, expressed this incentive in his book *Only the Paranoid Survive*, as did former Apple CEO Steve Jobs, who told his biographer, "If you don't cannibalize yourself, someone else will."[14]

An incumbent monopolist's incentive to innovate to preempt competition is the difference between its monopoly profit with the innovation and the profit that the firm would earn with the old technology when it faces an innovating rival. If the innovation is not drastic, it is more profitable for the incumbent to keep its monopoly by patenting the new technology than it is for a rival to patent the new technology and compete with the incumbent. The incumbent risks losing its monopoly profit if it does not innovate, while a successful new competitor can only profit from competition with the incumbent. The former monopoly profit is greater than the latter duopoly profit. Hence the monopolist has a greater incentive to invest to win a patent that would preempt competition. This turns on its head Arrow's finding that monopoly profits lower the incentive to invent.

This preemption argument should be interpreted with caution because it fails in some practical circumstances. First, if the innovation is drastic, a rival innovator would change places with the monopolist and wipe out the former monopolist's profits. Consequently, a new competitor has as much to gain from a drastic innovation as the former monopolist has to lose if it does not innovate and another firm succeeds. In this situation, the monopolist has no edge to prevent competition. The monopoly preemption incentive holds only if the innovation is not drastic. This result has important implications for the organizational theories of disruptive innovation that will be discussed in the next chapter. Claims that dominant firms are organizationally inept at responding to disruptive innovation overlook the fact that incumbent firms have no greater economic incentives to invest in disruptive change (i.e., drastic innovation) than do new competitors.

A second criticism is that preemption incentives do not necessarily extend to the more common situation of markets that are oligopolies, but not monopolies. Suppose that the incumbent shares the market with one or more other firms. In that case, it does not necessarily follow that the incumbent's profit from preventing the entry of a rival is greater than the rival's profit with the new technology when the innovation is not drastic. The reason is that a newly innovating entrant would take profit from all the incumbent firms, while an innovating incumbent shares the benefit from preempting new competition with other incumbents. These effects dilute the profit that any one incumbent can earn by preventing the entry of a new rival relative to the profit that an entrant would earn if it innovates. The incentive for preempting innovation by a new rival is strong only if the incumbent has a dominant position in the market that it can maintain by preventing entry.

Third, it does not necessarily follow that an incumbent would spend more to innovate than a new competitor if the incumbent cannot preempt rival innovation with certainty. Suppose there is a 50 percent probability that a rival will enter with a new technology and an equal probability that rival innovation will not occur. In the latter event, the incumbent's profit from its existing technology dulls its incentive to invest in R&D for an improved technology. This is the Arrow replacement effect. Although the replacement effect is operative only if rival innovation does not occur, the fact that there is a high probability that rival innovation will not occur reduces the incumbent's incentive to invest in the new technology. The reduction can be sufficient to make the monopolist's incentive to invest in the new technology less than a new competitor's incentive to invest.[15]

Fourth, preemption incentives do not necessarily foreclose a strategy in which rivals invest in R&D with the intention of selling or licensing discoveries to an established incumbent. An incumbent and a potential rival that owns a competing technology have incentives to avoid competition by negotiating an exclusive sale or licensing agreement.[16] If permitted by the antitrust laws, such ex post negotiation preserves the monopoly structure of production, with its adverse consequences for consumer prices. In that case, the incumbent monopolist would have no differential profit incentive to preempt rival innovation in order to preserve its monopoly. Many start-ups and research-focused organizations sell or license promising discoveries to established firms.

Fifth, and particularly important, a preemption strategy is unlikely to be profitable if it does not foreclose new competition. Spending R&D dollars to secure a patent does not foreclose competition if the incumbent's R&D efforts fail, or if potential rivals can invent around the patent and compete with a different technology. A preemption strategy can become an expensive game of Whack-a-Mole when potential rivals have multiple ways to enter and compete in an industry.[17] The constraint on the ability of a preemption strategy to foreclose competitors is important because it is rarely the case that acquisition of a single asset, such as a patent, is sufficient to protect a firm from future competition. A patent gives its owner a narrow right to exclude others from using the technology covered by the patent. It does not necessarily grant its owner a monopoly over a substantial region of economic activity.[18]

Nonetheless, it would be unwise to dismiss preemption incentives as a significant force in markets with established incumbents. The new-economy titans have won hundreds of bidding contests to acquire young start-ups, some of which have the potential (if only remote) to disrupt their dominance. Incumbents may win bidding wars because the acquisitions provide complementary benefits. In other circumstances, the incumbent may prevail because it has more to lose if a start-up matures to become a viable competitor than the start-up can expect to gain on its own by competing head-to-head with a dominant firm.

### Arrow does not consider interactions between a firm's product portfolio and the profitability of an innovation

Most innovations are new products, not simply new methods to lower a firm's production costs. Arrow's claims about relative payoffs from cost-reducing innovations in the monopoly and competitive regimes extend to product innovations if they are drastic (and if his other assumptions also apply, such as the assumption that the profit from an innovation does not depend on the identity of the inventor). However, Arrow's conclusions about monopoly and competitive innovation incentives do not necessarily hold for nondrastic product innovations, even if the monopolist is insulated from competition in R&D as well as in the product market.

A new product allows an incumbent monopolist to coordinate prices for the new product and for other products in its existing portfolio. This can increase the monopolist's profit from innovating relative to a new rival that can profit only from the innovation. In some situations, higher prices from this coordination deny consumers the benefits of greater innovation incentives, while in other situations, price coordination by the incumbent can promote innovation and consumer welfare.

Price coordination by an incumbent monopolist can increase R&D incentives and result in lower prices compared to innovation by a competitor if the innovation is a complement for the incumbent's existing products. The lower prices result from the Cournot complements effect discussed in chapter 2. The Cournot complements effect also implies that the incremental profit earned by a monopolist that makes a complementary innovation exceeds the profit that a new competitor can earn from the innovation. The monopolist also suffers a replacement effect that is absent for a competitor with no existing profits. If the replacement effect is not too large, consumers can benefit from greater innovation incentives and lower prices in the polar case of monopoly, compared to the polar case of innovation by a firm in a competitive market, when innovation involves complementary products.

It does not follow that monopoly is a superior organizational form for innovation when products are complements.[19] Competition in the supply of complementary products can create capabilities and incentives for innovation that are not available to a single firm. Consumers and firms that participate on the IBM-compatible PC platform benefit from competition that has generated thousands of complementary software and hardware innovations. Innovation in the field of plain-paper copiers flourished after a Federal Trade Commission (FTC) consent decree required the Xerox Corporation to license its patents, which opened the market to new competition for complementary innovations.

If the innovation is a substitute for the monopolist's existing products, a monopoly innovator can profit by coordinating higher prices for the new and existing products than would occur if a competitor innovates. Substitute products can be differentiated vertically or horizontally. Products with different qualities are vertically differentiated if, at equal prices, all consumers prefer a higher-quality product, such as a computer with a more powerful microprocessor. Products are horizontally differentiated if consumers prefer different products when they have the same price and quality. For example, some consumers have a brand preference for Ford trucks, while others prefer Chevy trucks.

Shane Greenstein and Garey Ramey analyze a model in which products are vertically differentiated.[20] In the simplest version of their model, a monopolist that is protected from competition earns the same profit from the new product as a firm in a perfectly competitive industry, provided that the new product is not a drastic innovation that eliminates demand for the old product. With minor changes in consumer preferences, the monopoly payoff from innovation can exceed the payoff to a firm in a competitive industry.

Yongmin Chen and Marius Schwartz examine product innovation incentives for a monopolist and competitive firms when products are horizontally differentiated.[21] In their model, a protected monopolist has a greater incentive to introduce a new product than does a firm in a competitive industry, again under the assumption that the new product is not a drastic innovation that eliminates demand for the old product.

In these models, the monopolist charges high prices for its products, which lowers the consumer benefits from the innovation. Monopolists can have greater incentives to invent than do competitive firms, but the incentives do not necessarily improve consumer welfare, after accounting for higher post-innovation prices. Nonetheless, consumers may benefit from monopoly innovation incentives if the innovations have large spillover benefits in other markets. These spillover benefits might not occur but for the greater innovation incentives of monopolies, and in some situations, they can more than compensate for the harm that comes from higher prices in the market where the innovation occurs.

A firm's ability to coordinate prices for a portfolio of products has an important message for innovation policy: One should not assume that the results in Arrow's model of innovation incentives for cost-reducing technologies apply without qualification to the more common situation of product innovation. Portfolio effects for multiproduct firms with monopoly power can create positive incentives for innovation that offset—and in some cases more than offset—the negative incentive from the replacement effect.

Durable goods create additional incentives for innovation. A durable good supplies services for its useful life. A washing machine does not need to be replaced until it wears out. A PC provides services until the hardware fails or the software becomes irreparably corrupted. In contrast, consumption exhausts the benefit from a nondurable good, such as gasoline, printer ink, or laundry detergent. Durable goods account for a large share of the growth of economic output and include many products that embody large investments in R&D.[22]

When a firm sells (rather than rents) a durable good to a customer, the firm can obtain no future revenue from the customer unless the product wears out or the customer has a desire to upgrade to a new version. If the good does not wear out and no new customers enter the market, the firm can sell more only if it lowers prices to attract more price-sensitive customers, or if it can convince existing customers to upgrade to a new version of the product.[23] The replacement effect is less significant for the seller of a durable good because the firm has no revenues that the innovation would eliminate, other than the revenues it could earn by lowering prices for existing customers or by attracting new customers. Chapter 6 offers an empirical example that illustrates the powerful effects of a durable good for innovation, even for a firm with monopoly power.[24]

Arrow's model ignores the dependence of appropriation on industry structure. He assumes that patent protection excludes imitation, and that an incumbent monopolist cannot obtain more value from an innovation than does a new competitor. An incumbent monopolist can have a greater incentive to innovate, even if it suffers the Arrow replacement effect, if the new competitor cannot match its ability to appropriate value from the innovation. The next chapter considers the interactions between limited appropriation and the Arrow replacement effect in a model of dynamic competition. The next chapter also explores the implications of managerial and organizational constraints for the relationship between competition and incentives to invest in R&D. Although innovation is not motivated purely by an economic comparison of expected costs and benefits, the power of economic incentives and disincentives for innovation, such as the Arrow replacement effect, should not be underestimated.

# 4 Dynamics, Cumulative Innovation, and Organizational Theories

What's the point of focusing on making the product even better when the only company you can take business from is yourself?
—Steve Jobs, "Voices of Innovation" (2004)

## 1 Introduction

Kenneth Arrow and Joseph Schumpeter are towering intellectuals associated with the theory of competition and innovation. Their contributions are profoundly important, but neither Arrow nor Schumpeter developed theories that describe innovation incentives that apply more generally to dynamic markets. Innovation is quintessentially a dynamic process. Furthermore, it is typically cumulative, with discoveries providing a knowledge base that enables future discoveries. Yet most theories, including the Arrow replacement effect, assume that innovation is a single discrete event with only limited consequences for the future evolution of industry technology and market structure.

Section 2 of this chapter reviews models of innovation incentives that embed both the Arrow replacement effect and Schumpeterian effects in dynamic settings. These models get complicated quickly and require simplifying assumptions to be analytically tractable. The section begins with a discussion of predictions from a model that has only two periods. Firms invest in research and development (R&D) in the first period to discover new products or new production technologies, and they choose prices in the second period for the products that emerge from the first-period R&D or compete using new production technologies.

A more dynamic formulation discussed in this section is a patent race in which investment in R&D yields a probability of discovery, and the first firm to make that discovery wins a patent. Both two-period models and patent races show that an increase in rivalry often, but not always, increases the probability of discovery. The two-period model shows that rivalry has an additional benefit: It increases expected future price competition when two or more firms develop innovative products that address similar market demand. The benefit of price competition is absent in models of patent races that assume that the patent excludes competition.

Industrial structure shapes the incentives for firms to invest in R&D and is shaped by the innovations that result from those investments. That interaction is absent in two-period models of R&D investment and in "winner-take-all" patent races. Section 2 also surveys dynamic models that allow this interdependence and explores their implications for the relationship between market structure, competition, and innovation. Models in which firms compete to improve technologies in discrete steps show that modest increases in competition can increase the probability of innovation, but intense competition can make innovation less likely. Competition increases the incentive to innovate to escape price constraints imposed by rivals, while also lowering the return to a firm that becomes a successful innovator by reducing the ability of the firm to profit from the innovation. Furthermore, intense competition can lower the probability that an industry is in a structural state that promotes R&D rivalry.

Many firms invest in R&D with the expectation that if their efforts show promise, they will be acquired by an incumbent or else license relevant technology to an established firm. When there is R&D for buyout, innovation incentives depend both on rivalry in R&D and on competition among incumbents that are likely bidders for the innovator or for licenses to its technology. Section 2 describes the implications of this bidding competition for R&D incentives.

Section 3 addresses economic incentives for firms to invest in innovations that build on prior innovations, also called "cumulative innovation." Cumulative innovation is a particularly important issue for the scope of intellectual property rights because strong patent rights provide incentives for early innovators, but potentially impose costs on firms that make infringing follow-up innovations. Xerography was a fundamental discovery that warranted patent protection. But plain-paper copiers were not widely accepted in the workplace until innovations in paper handling made these copiers an efficient office productivity tool. This section does not address the design of intellectual property rights. The focus instead is on antitrust policies that affect the exercise of these rights or other mechanisms that protect early innovators but have consequences for follow-on innovators.

Innovation is a human activity that often occurs in a complicated organization, which has limited information flows among the company's owners, managers, researchers, and customers, and may require structural adaptations to the existing business bureaucracy. Several theories of organizational behavior and bureaucracy address these limitations and their implications for innovation incentives. Section 4 reviews some of the more prominent theories and relates them to economic models.

## 2 Innovation Dynamics

We begin here with an extension of the discussion of incentives for innovation by exploring the implications from a model in which there are two separate periods. Firms invest in R&D during the first period, and successful innovators compete in price during the second period.

### Two-period models of R&D and price competition

This subsection summarizes key predictions from a two-period model presented in Gilbert (2019), which is closely related to models developed by Federico, Langus, and Valletti (2017, 2018) and Letina (2016). In the first period, firms can invest in any number of discrete R&D projects, each of which incurs a fixed cost and has an independent probability of success. For example, if each R&D project has a 50 percent probability of success, then if a firm invests in two projects, the probability that at least one succeeds is 75 percent. Innovation is winner-take-all if only one firm can profit from a discovery. Otherwise, firms can share profits from their discoveries, with competition reducing both the total available profit and each innovator's share of the total profit.

This relatively simple model has some useful implications. Suppose that firms have no profits from discovery that are put at risk and they cannot imitate discoveries made by others. If innovation is winner-take-all, then R&D incentives are independent of the structure of the market.[1] The reason for this result is that an additional R&D project has value only if all other R&D projects in the industry fail, without regard to whether these other projects are held by a single firm or distributed among many firms.[2]

A second implication from the model is that an increase in rivalry typically increases innovation incentives if innovation is drastic (as defined in chapter 3) but not "winner-take-all." If discovery is not winner-take-all, firms have incentives to innovate to take business from rival firms that have also innovated successfully. The opportunity to profit from simultaneous innovation provides a stimulus for R&D investment that increases with the number of potential innovators. Consequently, an increase in rivalry typically increases the total industry profit-maximizing level of R&D effort.[3] The opposite can occur if rivalry is intense. An additional rival lowers each firm's expected profit from R&D, which can more than offset its contribution to industry R&D efforts in some circumstances.

A third implication is that rivalry can reduce innovation incentives if innovations have only incremental benefits (i.e., if innovation is not drastic). Competition from existing products reduces the payoff from new products when innovations are not drastic. Consequently, industry investment in R&D for incremental innovations can peak at an intermediate level of market concentration.

A fourth implication is the importance of rivalry for future price competition. Rivalry can benefit consumers by increasing future price competition even if it lowers or has no effect on innovation incentives.

Fifth, the simple model also demonstrates that profits at risk from innovation reduce a company's incentives to invest in R&D (the Arrow replacement effect). An increase in the number of firms that invest in R&D can mitigate the replacement effect if competition among these firms lowers the profits that are at risk from innovation. A merger can increase the replacement effect and reduce R&D incentives if both firms have profits that are at risk from innovation, even if the firms' existing products do not compete with each other (see chapter 5).

Finally, information spillovers that allow rivals to imitate a discovery more easily can reduce incentives for firms to invest in R&D. It does not follow that such spillovers harm industry innovation. In the presence of information spillovers, an increase in the number of firms that can benefit from a discovery can lower each firm's expected profit from discovery and reduce incentives to invest in R&D; however, there is a compensating benefit because firms that benefit from information spillovers are more likely to innovate.

At a more general level, R&D competition promotes innovation by increasing the number of firms engaged in R&D, but R&D competition can harm innovation by reducing the ability of firms to appropriate value from their discoveries. Appropriation has several dimensions. An increase in competition lowers each firm's ability to profit from a discovery in the presence of other innovators. Furthermore, for innovations that are not drastic, competition from existing products can lower the profits from a discovery. The effect of competition on the value of an innovation is often called a Schumpeterian appropriation effect. An increase in competition also can reduce the ability of innovators to appropriate value from a discovery by increasing the number of potential rivals that can compete with the innovator by imitating the discovery.

Mergers can increase the ability of firms to capture value from discoveries by reducing the number of rivals that can profit from imitating a discovery. Furthermore, technological spillovers that occur within the walls of the merged firm can increase the value of an innovation or increase the efficiency with which the firm can conduct R&D. Chapter 5 describes these merger-specific issues in more detail.

### Patent races

One of the earliest approaches to adding dynamics to study the interaction between competition and innovation is the patent race. Firms compete by choosing R&D expenditures that generate a probability distribution for the discovery of a patentable invention. The first firm to realize the discovery wins the patent.

Simple models of patent races assume that patent protection excludes competition, so the winner of the race takes all or most of the available profit from an innovation. Patent protection is not essential for this property. Innovators in some industries can achieve temporary monopolies, or at least a very large increase in market share, in the absence of significant patent protection. The patent race model can be a reasonable description of R&D competition in markets without strong patent protection if firms have other ways to appropriate value, such as lead time, secrecy, or complementary factors of production.[4]

Generally, these models show that competition increases R&D effort and the probability of discovery, and reduces the expected time to patent an invention.[5] They also show that R&D competition can be socially excessive by incurring costs that exceed the benefits from the R&D, because firms ignore the negative effect of their R&D investments on the probability of discovery by others. This latter result has little policy significance for antitrust enforcement. The overwhelming empirical evidence is that R&D has large spillover benefits,[6] which suggests that, on average, firms invest too little rather than too much, in R&D.[7]

These early models are insightful and analytically convenient, but they offer only limited insight for strategic behavior that can have important consequences for competition and innovation. In particular, because these models assume that future success is independent of R&D that occurred in the past, they have no scope for strategic conduct that depends on a firm's technological lead compared to its rivals in the R&D competition.

For example, a firm that is first to make an intermediate discovery may gain an advantage that causes rivals to reduce their R&D efforts.[8] Under some conditions, the advantage can be so large that lagging firms cannot profitably leapfrog or even catch up to a firm that has made the intermediate discovery. In that case, lagging firms may drop out of the R&D competition entirely, leaving the technological leader as the only surviving investor. For firms that lag a technological leader, spending money on R&D can amount to pouring good money after bad if the leader is sure to maintain its technological edge. With its rivals out of the race, the leader can be content to invest in R&D at a profitable rate that can be much less than the industry would sustain if rival firms achieved technological parity.[9]

In other words, the effect of rivalry on R&D competition depends on the relative technological capabilities of the rivals, not just their number, which is a result that is confirmed in the empirical literature (see chapter 6).[10] R&D competition can be more intense in an industry with two firms that have similar technologies than in an industry with four firms if one firm has a substantial technological lead over its rivals. The next subsection describes a class of models that focus on innovation competition between firms that can occupy different positions in a technology space.

### Stepwise innovation

The stepwise innovation model emerged from collaborative efforts by Philippe Aghion, Christopher Harris, Peter Howitt, and John Vickers.[11] In their model, firms differ in the quality or cost of their product. Innovation is a discrete improvement in quality or cost, similar to climbing steps on a ladder. Firms can move higher up the ladder by investing to improve their qualities or lower their production costs, but first they must catch up to the current industry leader before they can become the next leader. These assumptions rule out innovations that allow a firm to leapfrog the current leader.

The stepwise innovation model illustrates incentives to invest to improve a product that depend on the product's position on the technological ladder relative to rivals' products. When competitors have similar qualities, there is an incentive to create a better product to escape competition and enjoy a temporary claim to higher profits. The greater the competition among firms that have technological parity, the greater is the incentive to innovate to escape this competition. This is the opposite side of the replacement effect in Arrow's static model. Greater competition lowers the profits from existing products. This implies a smaller replacement effect, and a corresponding greater incentive to create new products.

In Arrow's model, a successful innovator wins a patent that excludes competition for the life of the patent. Instead, the stepwise model plausibly assumes that a firm that catches up to a market leader has to share the profits from innovation with its equally efficient rivals, and may have its profits squeezed by competition from products that are lower down on the innovation ladder. Consequently, competition has a countervailing effect for innovation in the stepwise model. The greater the extent of competition in the market, the less a lagging firm can earn if it catches up to a market leader. Aghion and his colleagues characterize this as the *Schumpeterian effect*, in which competitive pressures dull innovation incentives for technological laggards.

Competition has countervailing effects for innovation in the stepwise model. Competition spurs firms to innovate, but competition also lowers the profit earned by a firm that innovates and catches up to the market leader. Some competition is desirable to stimulate innovation in the stepwise model. If products do not compete, there is little benefit from moving up the innovation ladder. The incentive to innovate also can be low if competition is intense because competition reduces the payoff from catching up to the market leader, although this has to be balanced against the stimulus from escaping competition.

The stepwise innovation model highlights the interdependence between industry structure, competition, and innovation incentives. An industry may appear to have little competition because firms have highly differentiated products, but that differentiation could be the result of innovation driven by intense competition. The effect of competition on the incentive to innovate differs for firms that are at the innovation frontier compared to those who are behind the frontier, and the overall effect of an increase in competition depends on the composition of firms in an industry, as measured by their technological prowess, which in turn is driven by innovation incentives.

The stepwise model demonstrates that intermediate levels of competition can best promote competition. This is the inverted-U relationship between competition and innovation. The stepwise model also demonstrates the importance of technological dissimilarities for innovation competition. Empirical researchers have attempted to verify the inverted-U relationship, but with limited success. Empirical support is more robust for the effect of technological capabilities on innovation. Several empirical studies show that competition has more pronounced benefits for innovation by firms that are technological leaders in an industry.

A limitation of the original stepwise model is its restriction to duopolies, with competition measured by the intensity of competition within each duopoly.[12] Antitrust authorities have limited scope to influence the degree of competition in an industry. They can prohibit certain exclusionary conduct and impose conditions such as compulsory licensing, supply obligations, or nondiscrimination obligations, but they cannot force firms to compete more or less.

Antitrust authorities can influence the number of firms that compete in an industry by choosing to block or allow a proposed merger. With the important caveat that a merger is not the same as an industry with one less rival, Gilbert, Riis, and Riis (2018) explore how results from the stepwise innovation model change when competition is measured by the number of industry rivals rather than the intensity of competition in a duopoly. In some circumstances, the rate of industry innovation reaches a peak level with a moderate number of rivals, while in other circumstances, greater rivalry increases the rate of industry innovation. These results depend on firm behavior, consumer demand, and the firms' technologies. The authors emphasize the importance of these factors for qualitative inferences about the effects of competition on industry innovation.

### Bidding for R&D assets

Many entrepreneurs invest in R&D with the intent to sell or license their results to an established company. For some, the strategy is a logical division of effort. Research organizations in the pharmaceutical and biotechnology industries specialize in the discovery of promising molecules or methodologies, and sell or license results to pharmaceutical companies that have the capabilities to administer and evaluate clinical trials, obtain regulatory approvals, and promote successful drugs. In the biotechnology sector, there were 68 acquisitions in 2014 alone, with a total reported value of $49 billion. In that year, biotech companies also entered into 152 licensing deals valued at an additional $47 billion.[13] For others, "innovate to sell" is the only practical strategy because established firms are protected by high barriers to competition and would crush upstarts if they try to compete. Facebook, Alphabet (Google), Amazon, and other tech goliaths have copied or threatened to copy promising new products.[14]

Competition in the acquiring market affects the incentive to innovate, in two ways. First, competition from existing products affects the maximum amount that a buyer would be willing to pay for an exclusive right to a new product. (I use the term "product" generically to describe any innovation, including R&D projects or new production technologies.) This maximum amount is the difference between the buyer's profit as the only supplier of the new product and the profit the firm can earn if it remains with an old product. I call the effect of competition on this profit difference the *value effect* of competition in the acquiring market.

A second effect arises from competition among potential buyers to acquire the innovation. Buyer competition for an exclusive right to a new product resembles an auction. Conditioned on the value of the new product, the winning bid is likely to be higher if more firms are interested in acquiring it. I call the effect of competition among buyers on the acquisition price the *auction effect*. For example, suppose that potential buyers of a new product are uncertain whether it is worth 80, 100, or something in between. With only two bidders, the acquisition price may be close to 80, while adding more bidders would increase the likelihood that the winning bid is closer to 100.

The auction effect is generally positive. The value effect is absent for a drastic innovation as defined in chapter 3 because competition from existing products does not affect the profit that a firm can earn from a drastic innovation. For a nondrastic innovation, the value effect is also positive if competition in the old product increases the profit that a firm can earn with the new product relative to what it would earn if the firm remained with the old product. In that case, both the value and auction effects imply that competition in the acquiring market increases the likely price paid to an innovator. Conversely, a merger of firms in the acquiring market lowers the likely acquisition price for an innovation and therefore lowers R&D incentives when the value effect is positive.

However, the value effect can be negative in some circumstances: An increase in competition in the acquiring market can lower the profit from a new product or technology relative to what a firm could earn if it remained with an existing product or technology.[15] In that case, the value and auction effects have opposite signs, and it is possible that the maximum return to an innovator would occur at an intermediate level of competition in the acquiring market. An implication is that a merger of firms in the acquiring market could increase the likely prices paid to acquire innovations and consequently increase R&D incentives when the value effect is negative.

Guillermo Marshall and Álvaro Parra describe innovation competition in a more detailed analysis that captures the auction and value effects for innovation incentives.[16] In their model, research laboratories engage in an innovate-to-sell strategy, while established firms compete with each other in the product market and with research labs in R&D. Consistent with the discussion in this section, they find that a decrease in the number of firms in the product market can lower the innovation rate if the value effect is negative, although that depends on an offsetting effect from the lessening of competition in R&D by firms in the product market that also compete in R&D.

## 3   Cumulative Innovation

The accumulation of knowledge enables innovation in the present and facilitates future innovations. In the words of Sir Isaac Newton, "If I have seen farther, it is by standing on the shoulders of giants."[17] Discoveries rarely spring de novo from the minds of inventors. Instead, they build on previous discoveries and add to the store of knowledge. The transistor and integrated circuit are typical examples of the evolution of cumulative innovation in the high-technology economy. The transistor implemented vacuum tube technology in a solid-state device. The integrated circuits that power numerous applications in the digital economy rely on the transistor and would not have been feasible without developments in photolithography to print circuit designs at microscopic dimensions.

Unfortunately, many important and influential theories of competition and innovation fail to address the trade-off between first-generation and subsequent innovation incentives. Arrow's replacement model treats innovation as a one-time event, as do simple models of patent races. In Schumpeter's theory of creative destruction, new innovations are waves that obliterate rather than build on prior discoveries. Rewards encourage innovation, but market rewards for discoveries can occur at the expense of rewards for innovations that build on those discoveries. Competition policies can balance these rewards, in part by preventing innovators from impeding subsequent innovators.

The history of many technological breakthroughs has dark chapters in which early inventors erected barriers to future innovation.[18] Wilbur and Orville Wright deserve accolades for their pioneering achievements in human flight. Among their accomplishments was the use of "wing warping" to steer an airplane, much as a bird bends its wings to change direction, for which they were awarded US Patent No. 821,393 in 1906. A few years later, Glenn Curtiss developed the use of ailerons to control flight, which is essentially the same technology used in modern aircraft. But the Wright brothers claimed that their patent also covered ailerons, and after years of litigation, they prevailed in court. Curtiss also won patents and resisted licensing new aviation entrants. The aircraft industry did not open itself to competition until 1917, when the US government compelled the industry to form a patent pool and license its patents. Also in response to concerns about the foreclosure of new technologies, in 1919, the government urged General Electric (GE), Marconi, AT&T, Telefunken, and Westinghouse to pool and cross-license their patents on radio technologies.[19]

Much of the debate about how to reward early and follow-on innovators emphasizes the design of intellectual property rights, which includes the scope of rights, their duration, and the novelty threshold for patentability.[20] Edmund Kitch argued that early inventors should have broad property rights that cover follow-on innovations, similar to the way that mineral rights for public land provide claimants with the exclusive right to prospect for additional discoveries within a defined region.[21] Broad and lengthy patents give more protection to initial inventors but, as a consequence of the Arrow replacement effect, they provide little incentive for follow-on innovations by the patentee that would replace its existing profits. A patentee can delegate subsequent innovation to others, but a broad patent scope allows the early innovator to require compensation from follow-on innovators because their innovations would infringe the original invention, absent a license. The compensation is a cost that reduces the profit from follow-on innovation.

Suzanne Scotchmer observed that, in general, it is not possible to design a system that provides efficient R&D incentives for both early and follow-on inventors without resorting to external subsidies, if they do not participate in an integrated R&D venture.[22] Greater rewards to one party necessarily occur at the expense of fewer rewards to the other party. Bargaining over licensing fees is problematic because licensing negotiations typically occur after firms have incurred R&D costs, which allows the early innovator to appropriate the benefits from these investments without adequately compensating follow-on innovators for their R&D expenses. Allowing innovators to bargain before R&D costs are sunk can solve the incentive problem, but that is a fiction if follow-on innovators are not known until they realize the results of their R&D investments.

Competition and intellectual property policies create trade-offs for incentives to invest in cumulative innovation. Potential inventors will not have financial incentives to incur costs to make a new discovery if they cannot anticipate receiving a healthy reward. However, rewards that take the form of exclusive rights can discourage future innovation by giving the first innovator the right to block future innovators or to impose a tax in the form of demands for large royalties in exchange for the right to use the first innovator's technology.

Figure 4.1 illustrates this trade-off for a frontier or first-generation invention that is also an input for follow-on innovation. The graph depicts a situation in which the first discovery can earn a per-unit royalty, which compensates the first inventor but is a cost for follow-on innovation. An example is the CRISPR technology for genome editing, which can be used to bioengineer new pharmaceuticals and agricultural crop protection products.[23]

In the graph, greater protection for first-generation innovation corresponds to a higher royalty. With perfect protection from imitation, the first-generation innovator can charge a royalty that maximizes its profit. Higher royalties, up to the profit-maximizing level $r^*$, encourage greater investment in the first-generation technology. However, higher royalties also lower the profit that follow-on innovators can earn. The royalty burden decreases the incentive to invest in follow-on technologies that require the right to use the innovation made in the first generation.

The optimal royalty strikes a balance between the reward for investment in first-generation innovation and the cost imposed for investment in follow-on innovation. Because the first-generation innovator achieves a maximum profit at $r^*$, a small reduction in the royalty below

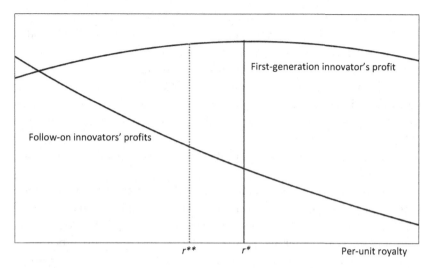

**Figure 4.1**
A small reduction in the royalty below $r^*$ increases the follow-on innovators' profits and has only a small effect on the first-generation innovator's profit.

$r^*$ would have little effect on the first-generation innovator's expected profit. The first-generation innovator's profit is almost independent of the royalty when it is close to $r^*$. In contrast, a small reduction in the royalty would have a large positive effect on profits and R&D investment for follow-on innovators. Consequently, total R&D investment for both first-generation and follow-on innovation would be higher at a royalty $r^{**}$, which is somewhat below $r^*$, than at the first-generation profit-maximizing level $r^*$. The total economic welfare also would be higher at $r^{**}$, for two reasons: Total investment would be higher and a lower royalty implies lower prices for follow-on innovations.[24]

A royalty that is below the level $r^*$ that maximizes the profits of the first-generation innovator can be achieved by patent policies that limit intellectual property protection for the first-generation technology, such as by narrowing the scope of follow-on inventions that would infringe the first-generation patent. Alternatively, antitrust policies can prohibit conduct by the first-generation inventor that may exclude or raise costs for subsequent innovators. These policies may include limiting the use of exclusive territories for practicing follow-on innovations or prohibiting contractual terms, such as obligations for follow-on innovators to grant licenses to their innovations for use by the first-generation innovator.

Policies that allow an original inventor to exclude competitors or raise their costs provide more compensation for early innovators, at a cost of lower compensation for new competitors.[25] In principle, antitrust enforcement should account for the cumulative nature of innovation and the interaction between current R&D investment and future innovation incentives, but in actuality, that is a tall order. It is hard enough to predict the consequences of a merger for current R&D investment decisions. It is even more difficult to predict the consequences of these decisions for the future evolution of an industry and future innovation incentives.

There is at least one respect in which the cumulative nature of innovation has clear implications for antitrust enforcement. This has to do with policies that affect the cost or access to existing technologies that interoperate with other technologies. Firms that develop innovative system products can exclude future innovators by designing their systems to be incompatible with complementary products. Antitrust enforcement has a role to play by facilitating access to existing technologies for follow-on innovators.

The US Department of Justice and the Federal Trade Commission settled some merger cases by requiring the parties to grant nonexclusive licenses that cover a broad range of technology. While compulsory licenses at noncompensatory rates may undermine innovation incentives for products that are easily copied, they can also facilitate innovation by reducing the cost of assembling necessary intellectual property rights, eliminating costly infringement litigation, and neutralizing efforts by firms that use intellectual property rights to block new competition.

A requirement to license intellectual property at low royalties can harm innovation by depressing incentives to create first-generation innovation by more than it promotes follow-on innovation, but that need not be the case. James Bessen and Eric Maskin develop a model of cumulative innovation in which discoveries have spillover benefits that promote future discoveries.[26] They find that innovation and consumer welfare can be higher if discoveries have no patent protection (which is equivalent to compulsory licensing at zero royalties) than in an alternative scenario in which discoveries have strong patent protection. Case studies discussed in chapter 7 also suggest that compulsory licensing can promote economic welfare in some circumstances.

## 4   Managerial Incentives and Organizational Failures

Up to this point, we have equated innovation incentives to the profits that a firm can earn by investing in R&D. These profits capture economic forces such as the Arrow replacement effect and Schumpeterian limitations on the ability of firms to appropriate the benefits from their innovations. But most R&D effort is exerted in hierarchical organizations by managers and staff who have compensation schedules and other motivations, such as security of employment, that are only indirectly related to the economic returns from innovation outcomes. In addition, cognitive biases can shape the R&D paths chosen by the managers of these organizations.

The next subsection reviews theories that address information flows within an organization to see how they complicate the relationship between competition and innovation incentives. While it is useful to review these theories, they do not converge on predictions that justify a substantial departure from the lessons learned from profit-based theories of innovation incentives.

### Managerial incentives

Economists have explored managerial incentives by treating the firm as a two-level hierarchy, with a principal (the firm owner) who has a claim on the firm's profits and an agent (the firm manager) who is paid by the owner and performs tasks on her or his behalf. There is a separation between ownership and control. With perfect information, the owner would simply direct the manager to perform the desired actions. In that case, the relationship between competition and innovation incentives does not depend on whether the owner of the firm performs R&D or whether she or he delegates that task to the manager.

In more realistic organizational settings, employees have better information than owners about the costs and payoffs from different projects, firm owners can only imperfectly monitor and enforce the activities that they would like their employees to do, and employees care more about their own welfare than owners' profits. Employees may want to work less, enjoy perks, avoid or take too much risk, or build empires that are unrelated to profitability. The owner's problem is to design a compensation schedule that incentivizes the manager to create value for the owner, but the separation of ownership and control implies that incentive schemes are imperfect in these situations. A key question is whether competition makes it easier or more difficult for owners to align managerial incentives with the owner's objectives.

Studies that have explored these agency theories reach different conclusions. Several theoretical studies find that competition reduces, but does not necessarily eliminate, the ability of managers to pursue their own objectives. Managerial choices that do not maximize profits increase the risk of bankruptcy, which harms managers as well as owners. Competition increases the risk of bankruptcy and should make managers less inclined to engage in unproductive activities.[27] Furthermore, when market outcomes are correlated, competition provides the owner (or owners) with information about firm performance, which she or he can use to benchmark managers.[28] These results do not imply that an increase in competition necessarily increases R&D incentives. Rather, they imply that the relationship between competition and innovation derived for owner-managed firms is more likely to hold for managerial firms when product markets are competitive.

Unfortunately, the concept that product market competition mitigates the problems caused by the separation of ownership and control is not robust to other reasonable assumptions.[29] Product market competition lowers profits, which squeezes the ability of owners to offer managerial compensation that induces desired efforts. If competition increases the risk of bankruptcy, with adverse consequences for employees, managers may avoid projects that have risky payoffs. Alternatively, managers may choose risky R&D projects if they profit handsomely when the project succeeds but bear only a fraction of the cost of failure.

It is useful to note that markets provide contrasting and generally ambiguous incentives for investment in risky R&D, even in the absence of complications from the separation of ownership and control. On the one hand, firms have incentives to invest in risky R&D projects because they profit most when they are the first to invent, but that has little social benefit if a second innovator would have soon followed.[30] This suggests too much incentive to invest in risky R&D. On the other hand, a firm, but not society, benefits when it innovates and takes sales from another innovator. This business-stealing effect is larger for incremental innovations, which implies a market bias in favor of low-risk projects and against risky R&D.[31]

Managerial theories of R&D investment have contrasting implications for R&D investment incentives that do not necessarily differ from the predictions of simpler economic models. For these reasons, I do not pursue managerial explanations for R&D investment further. A different theory, which has captured the attention of many corporate executives, stresses imperfections in the communication of information within an organization about the value of R&D decisions. This theory is the subject of the next subsection along with related theories of organization adjustment costs that affect innovation incentives.

### Organizational failures and adjustment costs

A large body of literature in the tradition of business strategy stresses how organizational limitations affect firms' R&D decisions. Clayton Christensen, a Professor at the Harvard Business School, sent shock waves through the world of corporate executives when he advanced the thesis that entrenched firms fail to make important innovations because they focus *too much* on the immediate needs of their customers.[32] According to Christensen, incumbents are not complacent firms that rest on their past accomplishments and ignore their customers' needs. Instead, he argued that incumbents are attentive to customers' desires to make their existing products better, but at the cost of missing the wave of disruptive new technology.[33]

Christensen's "innovator's dilemma" is that leading incumbents fail when faced by disruptive technologies because they continue to make the same kinds of decisions that made them successful. By focusing on the technologies that brought past success, they fail to encourage the adoption and use of new disruptive technologies, which ultimately leads to their demise. It is a type of managerial failure, although not one that results from a separation of ownership and control. Rather, it presumes that managers of established firms have beliefs based on years of industry experience reinforced by communication flows with customers and each other, and cognitive biases that cause them to focus on incremental improvements to existing technologies and to overlook potentially disruptive innovations. These new technologies begin as inferior alternatives to existing products but have the potential to improve and displace those products. When that happens, it is too late for the entrenched firms to catch up, and they are replaced by new market leaders.

Christensen supported his thesis with anecdotal evidence from industries such as computer hard disk drives (HDDs), where new generations of disk drives were sponsored by an industry outsider that ultimately displaced the former market leader. The main quality attributes of HDDs are the disk diameter (also called the form factor), the capacity, the durability, and the speed at which data can be recorded on and retrieved from the drive. Initially, smaller HDDs were inferior to larger ones. Established companies focused on making improvements to drives with existing form factors rather than pursuing these smaller, but initially inferior designs. New competitors focused on drives with smaller form factors, and they ultimately displaced incumbents as the industry gradually transitioned from 14-inch to 2.5-inch form factors.

There are many other examples of established firms that missed out on new technologies. Xerox PARC, a division of the Xerox Corporation, was responsible during the 1970s for important innovations such as the Ethernet, the graphical user interface (GUI) for personal computing, the computer mouse, and the laser printer. Xerox commercialized the laser printer but not these other innovations, presumably because the parent company was more focused on serving its existing base of office equipment customers.

Christensen's *The Innovator's Dilemma* is one of the best-selling business books of all time. The thesis is provocative and may explain the failure of leading firms to manage disruptive innovations in some circumstances. Yet, the theory does not explain why some companies fail in the face of destructive innovation, while others prosper. Some have questioned whether his theory is a good description of innovation patterns in many industries, including the hard disk drive industry.[34]

Moreover, there are alternative explanations for the failure of firms to catch the wave of creative destruction, which are based on more conventional economic thinking. The Arrow replacement effect tells us that earnings from existing products are a deterrent for innovation. For example, Polaroid did not lag in digital technology. Consistent with the science-driven philosophy of its founder, Edwin Land, the company invested heavily in R&D and held a strong patent position in digital imagery in the 1980s. But Polaroid was reluctant to give up a business model that produced a reliable profit flow from sales of instant film, which it would lose if it led a transition to digital imagery. By the time the company realized that digital cameras would replace instant film photography, the competitive tide had turned and Polaroid was left behind.[35]

Polaroid's demise is not a good example of a company that missed a technological opportunity because it was too focused on the immediate needs of its customers. Polaroid did not want to give up the profits that the company earned from its sales of instant film. Although organizational failure may have been a contributing factor, its conduct is consistent with the economic force of the Arrow replacement effect.

It is not necessary to call on organizational failures to explain why established firms often fail to catch the wave of creative destruction. The Arrow replacement effect explains why dominant firms may be reluctant to embrace innovation opportunities that would destroy their existing profits. In other circumstances, economic theory tells us that dominant firms can have incentives to employ preemptive strategies, such as aggressive patenting to avoid competition and maintain their monopoly profits. These preemption incentives apply only to incremental innovations and are absent for drastic innovations that would disrupt an industry.[36] Unless incumbents derive some unique benefit from a drastic innovation, or high barriers protect them from competition, it is not surprising that drastic innovations will emerge from new competitors, particularly in technological fields where R&D does not require highly specialized and expensive assets, thereby allowing many potential innovators.

Christensen's theory of demand-driven management failure also overlooks the positive role that incumbents can play as innovation incubators. Many successful new competitors are led by researchers and entrepreneurs who gained experience working for established companies. Intel was founded by engineers who worked at Fairchild Semiconductor International, which in turn was founded by dropouts from the Shockley Semiconductor Laboratory, a division of Beckman Instruments, and before that, Bell Laboratories. April Franco and Darren Filson demonstrate that the movement of experienced top executives from established firms contributed to the adoption of some of the more important innovations in the hard disk drive industry.[37]

Acquisitions also shaped the course of the industry. Seagate, one of the more successful suppliers of hard disk drives, purchased Control Data, Maxtor, and Conner, among others. Complementary assets were also a significant factor in Seagate's success. Seagate survived in part because it invested heavily in developing a supply chain that was capable of responding to the explosive growth of the computer market in the 1990s and 2000s.[38]

Rebecca Henderson and Kim Clark advanced an alternative organizational hypothesis to explain innovation successes and failures.[39] In contrast to Christensen's theory that attention to consumer demand causes dominant firms to miss out on disruptive innovations, Henderson and Clark focus on a firm's existing capabilities and how they match the requirements of new products or services. Organizational capabilities are costly to acquire and adjust. For Henderson and Clark, a radical innovation is one that requires very different organizational capabilities than the capabilities that a firm currently has. This supply-side definition of a *radical* innovation contrasts with Arrow's definition of a *drastic* innovation, which makes the technologies it replaces obsolete. According to Arrow, whether an innovation is drastic or incremental depends on consumer demand and the characteristics of new and old technologies. From the perspective of Henderson and Clark, whether an invention is radical or incremental depends on whether the innovation requires new technical and commercial skills or new approaches to problem-solving for the innovating firm.

Henderson and Clark's theory of radical innovation can explain why some dominant firms fail to innovate even in circumstances in which they have incentives to preempt potential rivals. For instance, RCA possessed the core capabilities to develop the first commercial portable transistor radio and demonstrated a prototype as early as 1952, but it was Sony, not RCA, that first commercialized the transistor radio—using technology licensed from RCA.[40] RCA had expertise in the components necessary to produce a commercial portable transistor radio, but innovation required the company to think differently about how the components could be integrated to produce a commercial product.[41] Intel curtailed research in reduced instruction set computing (RISC) microprocessors because they would not be backwardly compatible with its existing complex instruction set computing (CISC) microprocessor architecture.[42] New competitors, such as ARM, introduced RISC microprocessors for mobile technologies. The movement from CISC to RISC was a radical innovation given Intel's commitment to backward compatibility.

Rebecca Henderson tested alternative explanations for innovation using data on forty-nine photolithography projects undertaken by nineteen firms.[43] Semiconductor photolithography is an optical process used to transfer device and circuit patterns onto a semiconductor substrate for the manufacture of integrated circuits. She found support in the data for the economic theory that incumbent firms have incentives to invest in R&D to stay ahead of rivals when innovations are incremental and not radical in the organizational sense. However, for some innovations that involved radical organizational adjustments, incumbents failed to deter competition, even though they invested heavily in R&D for these new technologies.

Radical innovations do not have to doom incumbent firms. Canon and Fuji successfully navigated the transition from film to digital photography, while Kodak and Polaroid failed. Apple has successfully managed waves of potentially destructive innovations, perhaps because the company has not focused on consumer marketing research.[44] Other firms have attempted to minimize institutional inertia by isolating R&D teams from existing organizational distractions.[45]

## 5 Concluding Remarks

Innovation is fundamentally a human endeavor, but the power of economic incentives, even for the creative act of innovation, should not be dismissed. Arrow's replacement effect identifies an important economic force where profits that are put at risk from innovation create a drag on innovation incentives. Preemption incentives can turn this around in some circumstances.

Agency theories that account for the separation between corporate ownership and managerial control complicate the effects of profitability on firms' R&D decisions. Presently, there is insufficient convergence among these theories to reject more conventional profit-maximizing behavior. Organizational theories that emphasize cognitive limitations and the confines of existing corporate structures offer useful perspectives on the ability of firms to maintain technological leadership, but they often can be consistent with purely economic incentives for R&D investment.

A corporate decision not to pursue a new technological opportunity might be explained by information limitations or bureaucratic inertia. It is also a rational economic decision if the expected return from the new technology is not large enough to cover the cost of transitioning the company to the organizational structure that is more effective at supporting investment in the new technology. Demand-side cognitive influences (which Christiansen calls the "innovator's dilemma") suggest that established firms are more likely to pursue incremental than drastic innovations. Economic theories suggest that dominant firms can have incentives to invest more than rivals in R&D, but only for incremental innovations that do not make existing products obsolete. It is costly for organizations to adapt to radical innovations. Similarly, the Arrow replacement effect implies that profits from existing products that are at risk from innovation create opportunity costs for R&D. All of these theories imply that it is critical that established firms do not erect artificial barriers to new competition, because disruptive innovations are often likely to come from rivals that are new to the industry.

Competition policy for innovation should account for the types of innovations at issue, the technological capabilities of the merging firms, and the nature of competition in the affected industry. Although the next chapter shows that mergers can have adverse consequences for innovation incentives, a merger can promote innovation in some circumstances by allowing the merged firm to appropriate greater value from innovations or by facilitating technology transfer between firms with different technological capabilities. The ability of merger partners to benefit from their respective technological capabilities depends on whether innovations require organizations to adapt to radical change. Mergers are unlikely to have appropriation benefits for new technologies that are radical as defined by Henderson and Clark.

The so-called right model of innovation incentives depends on multiple factors, including market structures, technological opportunities, organizational capabilities, information spillovers, and other dimensions such as whether innovations are products or processes, or create durable goods. The details matter. A merger that reduces competition may dull incentives for innovation in one industry, and yet increase incentives for innovation or have no effect at all on another industry.

"We have made progress by finding partial theories.... The possibility that there is an infinite sequence of more and more refined theories is in agreement with our experience so far." This quote is from the eminent theoretical physicist and cosmologist, Stephen Hawking, writing in *The Theory of Everything*.[46] Although innovation economics is far removed from cosmology, his statement applies to theories of innovation incentives. Our existing models are partial theories that work in limited circumstances, and it is likely that more refined theories will follow.

The survey of the economics of innovation in this chapter and the preceding chapter may disappoint those who are looking for a general theory to guide antitrust enforcement for innovation. The empirical studies described in chapter 6 do not entirely resolve this dilemma because they find that competition often—but not always—promotes innovation. But first, in chapter 5, we turn to enforcement by antitrust authorities for mergers and acquisitions that affect innovation and future price competition.

# 5   Merger Policy for Innovation

Today, the goal of protecting innovation is often a decisive factor in our enforcement decisions involving technology companies, as well as in ensuing litigation.
—Renata Hesse, "At the Intersection of Antitrust and High-Tech" (2014)

## 1   Introduction

Chapters 3 and 4 discuss economic theories of innovation, and in particular the tension between the Arrow replacement effect, which identifies market power as a potential drag on incentives to innovate, and the appropriation effect associated with Joseph Schumpeter, which implies that size and market power can increase incentives to innovate by allowing a firm to capture greater value from its discoveries. This chapter describes how antitrust authorities have evaluated mergers that have consequences for innovation and competition for new goods and services, and considers how their enforcement decisions have aligned with lessons from the theory.[1] Chapter 7 reviews the success and limitations of remedies that US and European enforcement agencies have imposed to address innovation concerns for several mergers.

The US and other jurisdictions have generally similar approaches to evaluate mergers, including mergers that may harm innovation. They differ in the timing of their enforcement decisions and the road maps for appeals. These procedures can have significant consequences for enforcement decisions because they affect an antitrust authority's bargaining leverage. The parties to a merger may be more willing to negotiate a deal if they know that a challenge from the antitrust authority would require a lengthy appeal. I do not address these administrative distinctions. Even in the US, where the Department of Justice (DOJ) and the Federal Trade Commission (FTC) share merger authority, there are differences in procedures and appeals.

*Horizontal* mergers involve mergers (or acquisitions) of actual or potential competitors. The US antitrust agencies publish and update guidelines that explain their horizontal merger enforcement policies. I reference these guidelines in this chapter both as sources to explain current enforcement policies and as historical references that describe the evolution of merger policy for innovation.

*Vertical* mergers are mergers (or acquisitions) of firms that operate at different levels in a supply chain.[2] Vertical mergers can harm competition by foreclosing or making it costly for rivals to access inputs or customers. These mergers also can allow the parties to extend their market power into related activities or allow them to maintain their market power by requiring potential competitors to enter the supply chain at multiple levels.[3] Vertical mergers can promote innovation by allowing firms to coordinate innovation efforts and appropriate greater benefits from their results, but they can also create incentives to impede innovation by rivals. I discuss these issues in the related context of vertical integration in chapter 8 regarding Microsoft antitrust cases and in chapter 9 regarding antitrust investigations of Google's conduct in internet search.

My focus in this chapter is on horizontal mergers. Mergers can harm competition for products that firms currently sell and for new products. In addition, mergers can harm incentives for firms to develop new products or improve existing products.[4] For most of its history, antitrust policy focused only on the first type of harm: the likely consequences of mergers for competition in existing product markets. Although the antitrust laws have more general concerns, more than a century elapsed before courts and the antitrust enforcement agencies paid much attention to the two other types of harm from mergers: competition for new products and incentives to innovate.

These various competitive effects give rise to an enforcement matrix, with entries that describe the possible consequences from a merger in a high-tech industry. Combining possible price effects for existing and future products, the enforcement matrix has four entries, as shown in table 5.1.

Merger cases that fall in the southeast corner of the enforcement matrix ($P_{NH} + I_{NH}$) are not candidates for antitrust enforcement. These mergers raise neither concerns about higher prices nor harm to innovation. The northwest cell of the enforcement matrix ($P_H + I_H$) captures cases that raise concerns about both higher prices (for existing and/or

**Table 5.1**
The antitrust enforcement matrix for mergers in high-tech industries.

|  | Higher prices for existing or future products ($P_H$) | No higher prices for either existing or future products ($P_{NH}$) |
|---|---|---|
| Harm to innovation ($I_H$) | $P_H + I_H$ | $P_{NH} + I_H$ |
| No harm to innovation or an increase for innovation incentives ($I_{NH}$) | $P_H + I_{NH}$ | $P_{NH} + I_{NH}$ |

future products) and harm to innovation. Cases that fall in the northeast cell ($P_{NH} + I_H$) are pure innovation cases. These cases do not raise concerns about higher prices for either existing or future products, but they may harm consumers by reducing innovation incentives. Cases that occupy the southwest cell ($P_H + I_{NH}$) raise concerns about higher prices for existing or future products but could have offsetting benefits by increasing incentives for innovation. This can come about from appropriation effects that allow the merged company to achieve greater returns from research and development (R&D) efforts.

Merger authorities can clear a transaction, challenge the transaction unconditionally, or allow the transaction to proceed subject to remedies. Since 2010, the US antitrust agencies have reviewed an annual average of more than 1,500 proposed mergers and acquisitions under the reporting guidelines of the Hart-Scott-Rodino (HSR) Act.[5] The agencies investigate only a small percentage of the proposed transactions following a procedure called the "second request." In fiscal year 2017, the agencies reviewed more than 2,000 proposed mergers and acquisitions and initiated second-request investigations for fifty-one transactions, which resulted in thirty-nine challenges (about three-quarters of the transactions that received second requests). Most of these were settled with consent decrees in which the parties agreed to structural or behavioral commitments as conditions to complete the transactions. But in a few cases, the parties abandoned proposed transactions, or the agency pursued litigation. Fiscal year 2017 is notable for the large number of proposed mergers and acquisitions, but otherwise, the percentages of second requests, settlements, and court challenges are typical of other years.

Prior to 1995, merger challenges in the US and other jurisdictions rarely alleged harm to innovation. From 1990 to 1994, the DOJ and FTC filed complaints challenging a total of eighty-five proposed mergers. Ten of these proposed mergers were in high-tech industries such as computers and pharmaceuticals. The enforcement agency alleged harm to innovation in two of these cases (figure 5.1).

Allegations of harm to innovation in challenges to mergers in high-tech industries have become routine since the mid-1990s.[6] From 1995 to 2015, the agencies filed a total of 552 merger complaints, of which 144 were in high-tech industries. The agencies alleged harm to innovation in 124 of these cases—almost 90 percent of the complaints (figure 5.2).

Nearly all these challenges correspond to the northwest cell of the enforcement matrix ($P_H + I_H$). The antitrust authorities alleged that these mergers, if permitted without conditions, would raise prices (typically for existing products) *and* harm innovation. Deputy Assistant Attorney General Renata Hesse said the following about merger challenges that allege harm to competition in high-tech industries:[7]

Fortunately, we are rarely forced to choose between preventing higher prices and protecting innovation. Competition drives firms to compete on price and become more efficient, but it also can motivate them to invest more and work harder to improve their product design, function, and productive processes. In high-tech markets, a transaction that threatens to lead to higher prices or reduced output, therefore, will often have a corresponding negative effect on a firm's incentives to innovate.

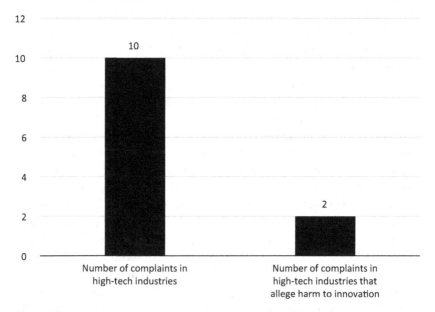

**Figure 5.1**
Merger complaints in high-tech industries: 1990–1994. Source: The data is from Gilbert and Greene (2015).

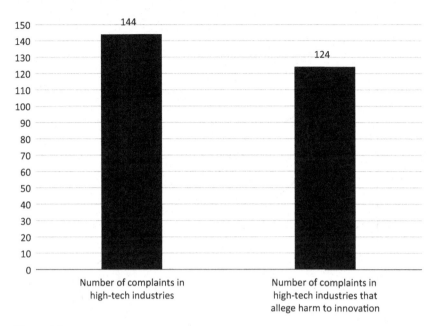

**Figure 5.2**
Merger complaints in high-tech industries: 1995–2015. Source: The data is from Gilbert and Greene (2015).

If concerns about harm to innovation and higher prices for existing products go hand-in-hand, then it raises the question whether there is much to be gained from adding innovation enforcement to the antitrust toolbox. Mergers that threaten harm to innovation would be challenged based on conventional concerns about higher prices.[8] But mergers can require enforcement agencies to weigh the risk of higher prices against possible benefits from greater innovation. Figure 5.3 illustrates such a merger. Line $D_1$ is the demand curve prior to the merger. It measures the amount that consumers, in the aggregate, would buy at every price. The premerger price is $P_1$. The merger increases the price to $P_2$ and increases the demand to $D_2$ by facilitating the creation of an improved product.

In the graph, the triangular area defined by the points $P_1AB$ is the consumer surplus prior to the merger—the difference between the amount that consumers are willing to pay for the good and the amount that they have to pay. The consumer surplus postmerger is the triangular area defined by the points $P_2CD$. The merger harms consumers in the aggregate if the area of $P_2CD$ is smaller than the area of $P_1AB$. Although that is a likely outcome based on theoretical results discussed

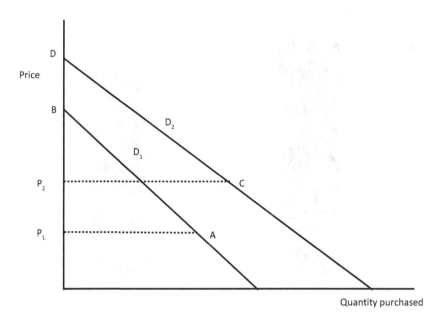

**Figure 5.3**
Demand and prices, with and without conduct that affects the price and quality of a product.

in chapter 4, it is not inevitable. The opposite could happen—the area of $P_2CD$ could be larger than the area of $P_1AB$.

The analysis of welfare effects from mergers that both increase prices and promote innovation (as illustrated in figure 5.3) is complicated because the net consumer harm or benefit depends on both prices and the shape of demand with and without the merger. Moreover, consumers may have different responses to price and innovation effects. Consumers with a high willingness to pay for an improved product may support a merger that raises prices and improves product quality. Other consumers could have little interest in the improved product. For these consumers, any significant price increase would lower their welfare.

This chapter begins in section 2 with a discussion of the Horizontal Merger Guidelines issued by the DOJ and FTC, a publication that describes how the agencies evaluate merger effects, and similar guidelines issued by the European Commission (EC). Most merger challenges that address innovation allege harm from unilateral effects, which are defined as effects that do not require accommodating responses from rivals. Section 3 examines the implications of market disruption and unilateral effects for the evaluation of mergers in technologically progressive industries. Section 4 describes three categories of mergers that raise possible concerns about unilateral effects for innovation: mergers that combine R&D projects and existing products; mergers that combine overlapping R&D projects; and mergers that combine overlapping R&D capabilities. Section 5 concludes by enumerating conditions that make a merger more likely to raise concerns about harm to innovation and future price competition.

## 2 Agency Guidance for Innovation

Agency guidance about the treatment of innovation concerns in merger cases was almost nonexistent until the most recent revision of the DOJ and FTC Horizontal Merger Guidelines, published in 2010. The first Merger Guidelines, published in 1968, addressed innovation only in the context of its effects on the market boundaries that the agencies employ to analyze price competition.[9] The next version, published in 1982 and revised in 1984, acknowledged that "sellers with market power also may eliminate rivalry on variables other than price"[10] but did not elaborate on that concept, other than to note that rapid technological change may complicate analysis of the effects of mergers on competition.[11]

The 2010 Horizontal Merger Guidelines are notable in a number of respects, one of which is their emphasis on innovation. They mention innovation no fewer than nineteen times, both in the context of potential anticompetitive harms from a lessening of innovation incentives and as a potential procompetitive benefit from an increase in the incentive to innovate. The guidelines include a separate section titled "Innovation and Product Variety," which notes that "competition often spurs firms to innovate" and states:

The Agencies may consider whether a merger is likely to diminish innovation competition by encouraging the merged firm to curtail its innovative efforts below the level that would prevail in the absence of the merger. That curtailment of innovation could take the form of reduced incentive to continue with an existing product-development effort or reduced incentive to initiate development of new products.

The guidelines further elaborate:

The first of these effects is most likely to occur if at least one of the merging firms is engaging in efforts to introduce new products that would capture substantial revenues from the other merging firm. The second, longer-run effect is most likely to occur if at least one of the merging firms has capabilities that are likely to lead it to develop new products in the future that would capture substantial revenues from the other merging firm. The Agencies therefore also consider whether a merger will diminish innovation competition by combining two of a very small number of firms with the strongest capabilities to successfully innovate in a specific direction.

The 2010 Horizontal Merger Guidelines acknowledge that mergers can increase innovation incentives by allowing the merged firm to appropriate more benefits from their discoveries:

The Agencies also consider the ability of the merged firm to appropriate a greater fraction of the benefits resulting from its innovations. Licensing and intellectual property conditions may be important to this enquiry, as they affect the ability of a firm to appropriate the benefits of its innovation.

The 2010 guidelines recognize efficiencies as a potential defense for an otherwise anticompetitive merger if they cannot be realized without the merger and if they are the type and magnitude that will benefit consumers.[12] It is likely that some R&D efficiencies, if they exist, are merger-specific. It is difficult to contract R&D to another firm, and collaborations that fall short of a merger can suffer from conflicts over the division of rights to inventions and a reluctance to share confidential information. While a merger does not guarantee that the parties will work smoothly to maximize the output of their R&D, it avoids some of the hazards that arise from interfirm R&D collaborations.[13] Furthermore, a merger may allow the merging parties to reposition their R&D assets to make expenditures more effective.[14]

Nonetheless, the 2010 Horizontal Merger Guidelines impose a high bar for efficiency defenses related to innovation:[15]

Efficiencies, such as those relating to research and development, are potentially substantial but are generally less susceptible to verification and may be the result of anticompetitive output reductions.

The agencies' cautious statements do not prevent approvals of transactions that promise significant innovation benefits. The Antitrust Division of the DOJ did not challenge a joint venture between Microsoft and Yahoo! that combined their back-end search advertising businesses. The Division concluded that the joint venture would create significant scale economies, which would enhance its ability to serve more relevant search results and paid search listings and enable more rapid innovation of potential new search-related products than would occur if Microsoft and Yahoo! were to remain separate.[16] The Division also did not challenge the merger of the mobile wireless companies T-Mobile and MetroPCS because its evaluation concluded that the combination would have a procompetitive impact on innovation and price competition by improving T-Mobile's scale and spectrum capacity.[17]

In these examples, the Antitrust Division did not have to choose between higher prices and innovation benefits because it either discounted the likelihood of higher prices or, as in the case of T-Mobile/MetroPCS, concluded that the combination would also promote price competition. While transactions may rarely require antitrust enforcers to choose between preventing higher prices and protecting innovation, Deputy Assistant Attorney General Hesse admitted that that can occur, and if it does, antitrust enforcers would have to weigh the consumer loss from higher prices against the gain from new or improved products or services, which would entail a complicated trade-off, such as the illustration in figure 5.3.

The Merger Guidelines issued by the EC make similar statements about possible innovation harms and benefits. The EC guidelines state: "Effective competition brings benefits to consumers, such as low prices, high-quality products, a wide selection of goods and services, and innovation," and also say that an increase in market power from a merger can allow the merging parties "to profitably increase prices, reduce output, choice or quality of goods and services, diminish innovation, or otherwise influence parameters of competition."[18] They specifically note concerns about a merger's effects on what they call "pipeline" products related to a specific product market, such as R&D projects that are at different stages of clinical testing, which is required before a drug can be approved for use.

While noting the potential harm to innovation from a merger, the EC guidelines also acknowledge that "a merger may increase the firms' ability and incentive to bring new innovations to the market and, thereby, the competitive pressure on rivals to innovate in that market."[19] The EC guidelines also note that efficiency claims must be verified and merger-specific; however, they do not add the qualification that efficiencies related to R&D are generally less susceptible to verification. The EC has vigorously pursued allegations of adverse innovation effects from a number of mergers, as demonstrated by the case discussions in chapter 7.

Some say the antitrust agencies and the courts give too little weight to efficiency claims,[20] while others allege that the agencies have been too quick to accept efficiency claims and have allowed mergers to be consummated that have harmed consumers.[21] Available evidence does not support overly aggressive enforcement by the DOJ and FTC against mergers that may affect the prices of existing products. A number of retrospective studies have identified price increases following mergers approved by the agencies.[22] Unfortunately, there is little research that addresses whether mergers have realized their claimed R&D efficiencies or innovation benefits for consumers. The case discussions in chapter 7 suggest that this is a much-needed area of study.

## 3   Evaluating Innovation Competition: Disruption and Unilateral Effects

Market structures and resulting competitive forces can create incentives for firms to influence the pace of innovation, while technical change can complicate merger analysis by transforming the markets in which firms compete, making it more difficult to predict future price and innovation effects.[23] While some argue that the disruptive effects of technical change demand a different approach to analyze mergers in high-tech markets,[24] the antitrust laws can accommodate the appropriate level of analysis without fundamental change.[25]

Antitrust agencies and the courts have taken anticipated market disruptions into account in their enforcement decisions. In a few cases, market dynamics have supported decisions not to intervene in mergers or other conduct.[26] In other cases, the mere fact that new competition may arrive at some point in the future has not been a justification for an exercise of market power before new competition occurs or a reduction in innovative effort that may delay or diminish the benefits of new technologies.[27] Many high-tech markets have characteristics such as economies of scale and network effects that erect barriers to new competition and can enhance the persistence of market power. It is particularly important that firms in these industries achieve market advantage as a result of open competition, and not from a merger that eliminates competition or from conduct that excludes rivals without compensating benefits.

Antitrust enforcers generally label the consequences of a lessening of competition as resulting from either coordinated interaction or a unilateral effect. Coordinated interaction involves conduct by multiple firms that is profitable for each of them only as a result of the accommodating reactions of the others, such as a price increase that is profitable only because rivals will increase their prices. Accommodation is not necessary for a profitable unilateral effect. In the merger context, a unilateral price (or innovation) effect arises if a merger makes it more profitable for the parties to raise prices (or lessen innovation) without accommodating responses by the parties' rivals.

Coordination requires either an explicit agreement or a mutual understanding to restrain from conduct that would be individually profitable without the agreement. Conspiracy is a coordinated effect that is enforced with an explicit agreement. The terms "tacit collusion" and "conscious parallelism" describe coordination without an explicit agreement.[28] Firms that tacitly collude recognize the interdependence of their actions. They enforce a collusive outcome with the threat that competition will break out and erode profits unless the participants adhere to the tacit arrangement.

R&D programs are often proprietary, their results are uncertain, and they occur in the distant future. Successful innovation can disrupt an industry and make rivals weak competitors, with little ability to impose costs on the innovator.[29] Taken together, these characteristics suggest that it would be difficult for firms to coordinate a reduction of innovation without an explicit agreement. Consequently, merger challenges that allege harm to innovation typically allege unilateral rather than coordinated effects.

Although allegations of unilateral competitive effects are more common in antitrust cases that involve innovation, several cases have alleged coordinated efforts to suppress innovation. The DOJ challenged a research joint venture in which major automobile manufacturers cooperated to develop pollution control technologies. The government alleged that the joint venture was actually a conspiracy to slow, rather than accelerate, progress because the technologies were costly to the manufacturers.[30] More recently, the EC fined truck manufacturers for colluding on the timing of compliance with emission control technologies[31] and private class actions alleged that Volkswagen AG—the manufacturer of Audi, Porsche, and Volkswagen automobiles— conspired with BMW, Mercedes-Benz, the parts supplier Robert Bosch, the automotive engineering company IAV, and the automobile association VDA (Verband der Automobilindustrie) to limit technological innovation and to share competitively sensitive commercial information about key aspects of the design and manufacture of their automobiles.[32]

Unilateral effects arise from a merger by eliminating competition that would otherwise exist between the merging parties if they remained independent. Suppose that an industry initially has three firms: Alpha, Beta, and Gamma. If the firms ignore the interdependence of their actions, each firm has an incentive to improve its product or innovate a new product to take business from the other two. Suppose that Alpha and Beta merge. Figure 5.4 illustrates the innovation incentives before and after the merger. Premerger, Alpha has an incentive to innovate to take business from Beta and Gamma, and Beta has an incentive to innovate to take business from Alpha and Gamma. The merged firm only has an incentive to take business from Gamma. Neither merging party benefits from sales that it takes from its merger partner.

The effects of a merger on the unilateral incentives for the merging firms to invest in R&D parallel the effects of a merger on the unilateral incentives of the merging firms to raise prices. Unilateral effects create "upward pricing pressure" from a merger that is proportional to the "diversion ratio" and the firms' profit margins. In the context of price competition, the diversion ratio measures the fraction of sales lost by one merging partner from a price increase that would be captured by the other merger partner. A merger internalizes these lost sales and a firm's profit margin measures the benefits from capturing the business.[33] Similarly, as illustrated in figure 5.4, before the merger, each firm has an incentive to invest in R&D to divert future business from its rivals, including its merger party. Joseph Farrell and Carl Shapiro define an "innovation diversion ratio" between Firms A and B that is equal to the fraction of additional gross profits earned by Firm A when

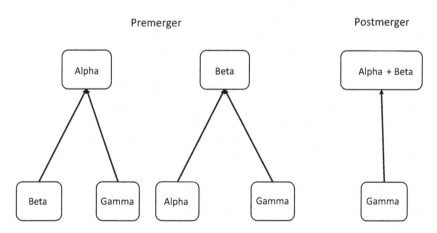

**Figure 5.4**
Premerger, Alpha and Beta have incentives to take business from Gamma and from each other. If Alpha and Beta merge, the merged firm only has an incentive to take business from Gamma.

it devotes more resources to innovation that come at the expense of Firm B.[34]

The unilateral price effects from a merger understate the effects of the merger if rivals raise their prices in response to higher prices charged by the merging parties. That reaction would follow if pricing decisions are "strategic complements," which they are in many business situations.[35] In that case, the existence of unilateral upper pricing pressure from a merger is an indication that the merger will cause generally higher industry prices if there are no offsetting efficiencies. However, there is no a priori reason to expect that R&D decisions are strategic complements. Rivals could profitably increase R&D expenditures in response to a reduction in R&D effort by the merging parties. That is, R&D decisions may be "strategic substitutes."[36] If R&D decisions are strategic substitutes, unilateral downward innovation pressure from a merger is not a reliable indicator that the merger will more generally harm total industry investment in R&D and innovation.

Several recent theoretical studies illustrate downward innovation pressure from mergers.[37] Nonetheless, a merger can promote innovation if the merging parties can conduct R&D more efficiently, or if the merger stimulates rivals to increase their R&D investment. Downward innovation pressure from a merger can also be reversed for reasons related to appropriation, technological spillovers, and innovation dynamics that simple models do not capture.[38] Furthermore, if discovery is "winner-take-all," a merger would have no effect on innovation incentives because the benefit from making a second innovation is zero, whether the innovator is an independent company or a party to a merger.[39] Of course, a merger can still harm consumers by eliminating price competition that would have occurred without it, even if it has no adverse effect on innovation.

A merger can enhance the ability of the merging firms to appropriate value from their discoveries in different ways. A merger can reduce future product market competition, and the expectation of higher future prices can be an inducement for the merged company to increase investment in R&D. Consumers may benefit from the resulting increase in innovation, but that can be more than offset by the penalty from higher prices.

Appropriability is also related to the presence or absence of technological spillovers. A merger can benefit appropriation by reducing interfirm spillovers or by exploiting intrafirm spillovers that occur by transferring relevant knowledge between merging parties. Interfirm technological spillover measures the extent to which a discovery benefits one or more rival firms by making it easier for rivals to imitate the discovery. A merger can increase the incentives of the merging parties to invest in R&D by eliminating a rival that can benefit from interfirm spillovers. That increase does not necessarily imply greater industry innovation. Recall from the discussion in chapters 3 and 4 that technological spillovers can increase total industry innovation by making it easier for rivals to imitate discoveries. For this reason, merging parties should have a high burden to prove that a reduction of interfirm spillovers is a cognizable merger-specific benefit. An exception is an extreme case in which the risk of imitation is so high that the merging parties would not profitably invest in R&D, and the merger sufficiently reduces this risk to make postmerger R&D investment profitable.

Intrafirm technological spillover measures the extent to which a party to a merger benefits from a discovery made by the other party (or more generally, technological spillovers between different divisions of the same firm). A merger can increase appropriation by promoting intrafirm spillovers.[40] A simple example, similar to the example in chapter 3, illustrates the critical role of intrafirm spillovers for the effects of mergers on innovation incentives. Suppose that three firms sell an identical product at a price of $10. Each firm initially has a constant marginal cost of $8 and sells 1 million units, yielding a gross profit of $2 million. The firms can invest in R&D to lower their costs by $2. If the price and sales are unchanged, the lower cost adds $2 million to a firm's gross profit. The firms can benefit from the invention only to the extent that it lowers their own costs; they cannot profitably license the invention to others.

Now suppose that two firms merge. First, assume that the merger does not change the price or the merging parties' sales. Each party continues to sell 1 million units at a price of $10. If there is no intrafirm spillover, any cost reduction applies only to the sales made by the party that invests in R&D. Under these assumptions, the merger would have no effect on innovation incentives.

Suppose instead that there are large intrafirm spillovers: The merged firm can apply a cost reduction to its entire output of 2 million units. In this case, the merger doubles the value of cost-reducing R&D. A $2 cost reduction by the merged firm increases its gross profit by $4 million, whereas prior to the merger, a $2 cost reduction would increase profit by only $2 million. The larger postmerger base of sales would give the merged firm a greater incentive to invest to lower its costs.

Thus, whether the merger increases or decreases innovation incentives depends critically on whether investments are specific to the output of each merging firm or whether intrafirm spillovers allow investments to benefit both of the merging firms. The example assumes that the merger does not change prices. If the merged firm raised prices, demand would fall, as would the output that can benefit from cost-reducing R&D.[41]

Intrafirm spillovers are also relevant for product innovations. Suppose that a company that sells heavy-duty automatic transmissions merges with a company that sells light-duty automatic transmissions. There are strong intrafirm spillovers if a transmission improvement discovered by one of the merger parties would be easily adopted and incorporated into products sold by the other party.

Absent efficiency and appropriation benefits, simple models of mergers between otherwise similar firms imply a reduction in R&D incentives from downward innovation pressure. These results contrast with predictions from some of the dynamic models described in chapter 4, which show that a reduction in competition can increase as well as decrease innovation incentives (e.g., the inverted-U relationship between competition and innovation described in papers by Phillippe Aghion and the coauthors).[42] Some attribute these differences to the fact that these dynamic models study the effects on innovation incentives from a change in competition; they do not study the effects of a merger. A merger differs from an industry with one less rival because the merger, at least in the short run, does not eliminate the R&D capacities of the merging firms; instead, it allows the merged firm to coordinate R&D and other decisions for the merging parties.

This distinction between a merger and a reduction in competition is significant, leading some to argue that the theoretical literature on competition and innovation is misleading.[43] However, it is important to appreciate that recent theoretical models of downward innovation pressure from a merger are highly simplified and omit crucial interactions between market structure and innovation incentives that are more carefully considered in dynamic models of competition and innovation. Furthermore, with the passage of time, the distinction between a merger and a reduction in competition tends to evaporate as the merging firm repositions its R&D assets, making the technological capabilities of the merged firm more similar to its industry rivals.

One insight from dynamic models is that the effects of rivalry on incentives to innovate depend on firms' relative technological positions as measured by their distance to the technological frontier practiced by the most advanced firms in the industry. Firms that are behind the frontier have little incentive to catch up to a market leader if the lagging firm cannot appropriate significant value from an innovation. In contrast, firms that share a position of technological parity have large incentives to innovate if the market is highly competitive because innovating allows a firm to escape competitive discipline. Innovation incentives also depend on whether a future discovery is drastic (meaning that it makes alternative technologies noncompetitive), radical (meaning that it requires organizational adaptation), or incremental, as well as on other factors, including the extent of intrafirm technological spillovers.

Simple models that examine the effects of mergers on incentives to invest in R&D ignore or do not fully address these complications. It is a valid criticism that existing dynamic models of competition and innovation do not capture the downward innovation pressure from a merger caused by the internalization of diversion effects between the merging parties, but it is premature to dismiss the relevance of dynamic models of competition and innovation for insights about the effects of mergers for innovation incentives.

## 4   Three Types of Mergers That Can Have Unilateral Effects for Innovation

US antitrust authorities have alleged harm to innovation in almost all their challenges to mergers in high-tech industries. Most of these challenges address harm to competition in existing product markets, and the innovation allegations are ancillary to the product market complaints. In other cases, the innovation allegations stand on their own and, for cases resolved with consent decrees, determine the scope of negotiated remedies. These innovation-centric merger cases involve alleged unilateral effects for innovation that fall into three general classifications. The classifications are not mutually exclusive because mergers can raise concerns about innovation and future price competition that correspond to more than one category.

### Product-to-project mergers
A "product-to-project" merger involves a firm with an existing product that merges with, acquires, or is acquired by another firm that has an R&D project which, if successful, would compete with the other firm's product. An example of this is a merger between a firm that has an

existing hepatitis vaccine and another firm that has a drug in clinical trials that may prove to be a safe and effective hepatitis vaccine.

Product-to-project mergers can harm future price competition if the project is successful and would compete with the other merging party's product. Furthermore, product-to-project mergers can lower innovation incentives because they create an Arrow replacement effect. The project, if it is successful and competitive with the other merging party's product, would erode profits from the existing product. This replacement effect lowers the merged firm's total expected profit from the project and the existing product compared to the combined profits for the firms as independent entities, and can lead the merged firm to terminate the project or delay its progress.

The Arrow replacement effect is a type of unilateral business-stealing effect that operates within the walls of the merged firm, because a successful project would steal business from the merger partner. The competitive significance of the replacement effect depends on the extent to which a merger increases the profits that are at risk from innovation relative to the profit that an independent firm can earn by innovating. A merger causes a replacement effect if either merging party has an existing product whose profits are at risk from a discovery by the other party. The merger can increase premerger replacement effects, and create additional downward innovation pressure, if both merging parties have products whose profits are at risk from discovery by the other merger party. If the products are substitutes, the merger could increase the replacement effect by increasing prices and the profits at risk from a discovery. In that case, concerns about innovation would go hand-in-hand with conventional concerns about price effects in existing markets. A merger can also increase the profits at risk from discovery if the merging parties' existing products are not substitutes for each other. The harm from this potential replacement effect is not present in merger evaluations that focus on prices.

For example, suppose that two firms are engaged in R&D which, if successful, would create a new oral form of insulin that would displace insulin administered by injection. One of the firms currently sells an injectable insulin. The other firm sells a pump that is used to administer the injectable medication. If they merge, conventional analysis would recognize that the injectable drug and the pump are complementary products, and that placing them under common ownership could be a merger benefit. The Cournot complements effect suggests that the merged firm would profitably coordinate lower prices for the injectable drug and the pump and might have improved incentives to promote their use.

Ignoring innovation effects, the benefits from complementary products would support a decision to approve the merger. Nonetheless, the merger could greatly increase replacement effects and consequently reduce the merged firm's incentive to invest in R&D to create an oral insulin drug.[44] The drug, if successful, would eliminate profits from both the injectable drug and the pump used to administer that drug. Although these products are complements and not substitutes for each other, profits from both of these products would be at risk of being eliminated by the discovery of an effective oral insulin medication.

Observe that price and innovation effects from this example act in different directions. The merger can lower prices for the existing injectable drug and pump because the products are complements. However, the merger can lower incentives to invest in an oral insulin drug because it increases the replacement effect.

### Project-to-project mergers

"Project-to-project" mergers involve firms that are both actively involved in R&D directed toward similar applications. An example of this is a merger of firms in which both firms have potential hepatitis vaccines in clinical trials. Another is a merger of consumer research companies, each of which is testing a new way to collect data on consumer viewing patterns, such as the merger of Nielsen and Arbitron discussed in chapter 7.

As in product-to-project mergers, project-to-project mergers can raise antitrust concerns because the merger can cause the merging parties to terminate or delay projects that target similar applications. In addition, project-to-project mergers can raise antitrust concerns related to potential future price competition, but with the additional complication that these future markets do not presently exist. Project-to-project mergers threaten future price competition only if both of the merging parties' projects would have been commercially successful without the merger and would have competed in the same market. In such a merger, a project is a potential competitor for the project owned by a merger partner, but only if the partner's project is commercially successful.

Project-to-project mergers can cause one of the parties to abandon a project, or cause one or both of the parties to invest less and move a project more slowly through R&D. Project-to-project mergers also can have efficiency benefits by allowing the merging parties to benefit from complementary skills, or by exploiting knowledge transfers (intrafirm technological spillovers).

## Overlapping R&D mergers

A third type of merger with potential unilateral effects for innovation is a merger of firms with R&D capabilities that can be applied to similar applications but have no identified projects that are directed to these applications. An example is the merger of Ciba-Geigy and Sandoz discussed in chapter 7. At the time of the merger, each company had established R&D capabilities and owned intellectual property rights for gene therapies, but neither firm had a gene therapy product or even a gene therapy project in a promising state of development.

Mergers of firms with overlapping R&D capabilities can have harmful effects for innovation and future competition. The evidentiary hurdles are larger for these types of mergers because, lacking information about projects or products, it is uncertain whether the merging parties' R&D efforts will be directed toward similar ends and, if they are successful, whether they will result in products that compete with each other. Furthermore, as in project-to-project mergers, the integration of R&D capabilities can have efficiency benefits, and the merger can increase innovation incentives by exploiting intrafirm knowledge transfers.

As in any enforcement decision, antitrust authorities that evaluate the possible effects of a merger on innovation and future price competition must navigate errors of overenforcement and underenforcement. Some types of mergers are likely to allow a relatively precise estimate of potential harm. For product-to-project mergers, an antitrust authority can identify R&D projects and the products whose profits are at risk if the projects are successful. This information allows a confident estimate of the Arrow replacement effect and a relatively precise estimate of harm from the elimination of future price competition if the R&D projects of one or more of the merged firms are successful. These considerations, taken together, suggest a relatively high degree of certainty if the investigation of a product-to-project merger indicates that the merger would be anticompetitive.

Information about the competitive effects from project-to-project mergers is somewhat less precise. These transactions can cause the merged company to delay or terminate R&D projects, but they can also have efficiencies from complementary activities and knowledge transfers between the merging parties. Furthermore, future price effects are more uncertain than for product-to-project mergers because they occur only if both merging parties have successful projects that yield products or technologies that are substitutes for each other. There is no loss of future price competition if only one of the merging parties has a successful project or if both merging parties have successful projects but the resulting products or technologies do not compete with each other.

Estimates of the effects of mergers that combine overlapping R&D capabilities on innovation and future price competition are likely to be less precise than estimates of effects for either product-to-project or project-to-project mergers. Investigations may not reliably predict whether the merging firms would have directed their R&D capabilities toward similar ends if they had not merged. If they would have had done so, their efforts may not have been successful, and, if they were both successful, it would be difficult to reliably predict whether the resulting products or technologies would have been substitutes for each other. Moreover, mergers of overlapping R&D activities can have significant efficiency benefits from combining complementary assets and intrafirm knowledge transfers.

These considerations give rise to an enforcement pyramid for mergers that may threaten innovation (see figure 5.5). Moving up the pyramid, the benefits from a merger challenge become more uncertain, as does the evidentiary burden for proof of harm in litigation. The slices of the pyramid are indicative of the relative number of cases in which alleged innovation harms were central to enforcement outcomes. Many cases have alleged product-to-project competition. A smaller number have alleged project-to-project competition, and only a few cases have alleged harm from overlapping R&D capabilities without identifying specific projects or products.

Mergers that involve product-to-project competition are analogous to potential competition cases, in which a firm in an existing market acquires another firm that is a potential entrant in its market. US courts have resisted allegations of harm from eliminating a potential competitor.[45] A major evidentiary stumbling block is the requirement of proof that the acquired firm would have entered the market.[46] Similar evidentiary concerns apply to product-to-project innovation cases, but not to the same extent. Information about projects in a R&D pipeline facilitates the identification of potential competitors and helps to establish the likelihood that projects will advance to commercial viability.

R&D pipelines for some products can be long, and projects in the pipelines can have a low probability of success.[47] During the time

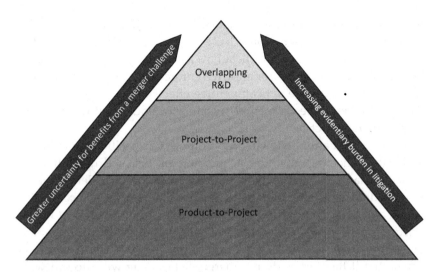

**Figure 5.5**
The enforcement pyramid for mergers with potential harm to innovation.

required to evaluate a promising drug compound, other therapies may emerge from unexpected sources, the compound may fail, or it may succeed in an application that was not predicted.[48] However, the fact that a project is at an early stage of development and success is uncertain should not make acquisition of the project immune from antitrust enforcement. The acquisition should raise antitrust concerns if there is a nonnegligible probability that the project would have succeeded without the acquisition and, conditional on success, the acquisition would have anticompetitive effects.

For example, suppose that Alpha, a pharmaceutical company with a dominant position in drugs for the treatment of thyroid cancer, seeks to acquire Beta, a company with a thyroid cancer drug in its R&D pipeline. The reviewing antitrust authority is concerned that, postacquisition, Alpha would delay or terminate clinical trials for Beta's drug.

Does it matter whether, if Beta remains independent, the odds of success for Beta's drug are 50 percent, 5 percent, or a different number? If there are no efficiency justifications for the acquisition, and if antitrust enforcement has no opportunity costs, the probability of success should not determine the enforcement decision. There are two scenarios to consider that correspond to the success or failure of Beta's project if Alpha does not acquire Beta. In the success scenario, the acquisition could harm patients by terminating or delaying Beta's project, or by reducing price competition that could occur if Beta remains independent. The acquisition would have no anticompetitive effect in the alternative scenario, in which Beta fails as an independent competitor. The substantive questions for antitrust enforcement are (1) whether the acquisition would have anticompetitive effects in the success scenario and (2) whether there are merger-specific efficiencies that would offset these adverse effects.

Suppose that after reviewing the proposed acquisition, an antitrust authority concludes that the acquisition of Beta's drug will have expected harm $H$ compared to the outcome in which Beta remains independent, assuming that the project would have been successful if Alpha and Beta did not merge. The expected harm is the loss of future price competition and the cost of any delay in Beta's drug or reduction in its probability of success as a result of the merger.[49] The agency estimates that if the parties do not merge, Beta's drug will succeed with probability $p$. The acquisition also may have efficiencies that are specific to the acquisition that can be realized if Beta's project fails. The agency estimates that these merger-specific efficiencies have an expected benefit $Z$.

The reviewing agency should challenge this acquisition if the net expected harm is positive. Assuming that the acquisition does not change the success probability, the net expected harm is

$$pH - (1-p)Z.$$

The first term is the expected harm from the acquisition of a project: the harm from the acquisition multiplied by the probability that Beta would have developed a safe and effective drug if it had not been acquired. The second term measures the expected merger-specific benefit from the acquisition given that the project would have failed if it had not been acquired. If $Z = 0$ (i.e., there are no merger-specific efficiencies if Beta's project would have failed without the acquisition), then the agency should challenge the acquisition if $H > 0$; that is, if the acquisition would harm innovation or future price competition assuming that it would have been successful without the merger.[50] This estimation of harm does not depend on the project's success probability if Beta remains independent. Of course, an antitrust authority may choose not to challenge an acquisition for which $H$ is positive if the expected harm, $pH$, is less than the enforcement cost or the opportunity cost of foregoing other enforcement actions. For example, suppose that the reviewing antitrust agency estimates that the acquisition will increase consumer costs by about $10 million compared to the scenario in which an independent Beta would develop a safe and effective drug. The agency also concludes that the probability that Beta would be successful on its own is only about ten percent. The expected harm of about $1 million may not be large enough to warrant action by the agency given its other enforcement opportunities.

If the acquisition has the potential to harm innovation or future price competition, merger-specific efficiencies can be a credible defense, either by negating the net harm, $H$, from a successful project or from complementary assets that do not depend on the project's success (i.e., the amount $Z$). A review of the transaction can determine whether the parties have benefits from complementary assets that do not depend on the success of the acquired drug project and whether the acquisition is necessary to realize these benefits.

For project-to-project mergers and mergers that combine overlapping R&D capabilities, a key question is whether structural presumptions should apply to identify possible harm to innovation from mergers that increase the concentration of R&D activity. The Supreme Court endorsed a presumption that increases in market concentration cause higher prices in its 1963 *Philadelphia National Bank* decision,[51] and the Horizontal Merger Guidelines published by the DOJ and FTC and the EC guidelines include rebuttable structural presumptions for price effects.

The Antitrust Guidelines for the Licensing of Intellectual Property, published by the US DOJ and FTC in 1995, introduced the concept of an "innovation market," subsequently renamed a "research and development market" when the agencies revised the intellectual property (IP) guidelines in 2017. The determination of a research and development market can be a first step to evaluate the concentration of R&D activity and how it may be affected by a merger or other conduct. According to the IP guidelines:[52]

A research and development market consists of the assets comprising research and development related to the identification of a commercializable product, or directed to particular new or improved goods or processes, and the close substitutes for that research and development. When research and development is directed to particular new or improved goods or processes, the close substitutes may include research and development efforts, technologies, and goods that significantly constrain the exercise of market power with respect to the relevant research and development, for example by limiting the ability and incentive of a hypothetical monopolist to reduce the pace of research and development.

Antitrust scholars and practitioners have criticized the concept of an innovation or R&D market.[53] R&D does not typically trade in a market; R&D is an input to innovation, not the output of innovation; and the relationship between innovation and a concentration of R&D effort is complex. Nonetheless, the logic of unilateral effects demonstrates that under some conditions, a merger can reduce incentives to invest in R&D for much the same reason that a merger would result in higher prices, although other factors, such as appropriability, that are not present for price effects can reverse the harm to innovation from a merger.

While there is no definitive connection between concentration in a research and development market and innovation, this does not negate the utility of defining a research and development market as a tool to help identify firms that have the ability and incentive to develop new or improved goods or processes directed toward a commercial product. All else being equal, consumer harm from a project-to-project merger or a merger of firms with overlapping R&D capabilities is more likely if the merging firms participate in a relevant research and development market that is highly concentrated. Furthermore, a research and development market can be used as a screen to identify situations in which a merger is unlikely to harm innovation, just as price effects are unlikely in unconcentrated markets.

Antitrust authorities have supported structural screens in research and development markets for which harm to innovation is unlikely. Regarding potential competitive effects from an intellectual property licensing arrangement in a research and development market, the DOJ and FTC Antitrust Guidelines for the Licensing of Intellectual Property state that:[54]

The Agencies, absent extraordinary circumstances, will not challenge a restraint in an intellectual property licensing arrangement if (1) the restraint is not facially anticompetitive and (2) four or more independently controlled entities in addition to the parties to the licensing arrangement possess the required specialized assets or characteristics and the incentive to engage in research and development that is a close substitute of the research and development activities of the parties to the licensing agreement.

The agencies' Antitrust Guidelines for Collaboration among Competitors go further, stating that "absent extraordinary circumstances, the Agencies do not challenge a competitor collaboration on the basis of effects on competition in an innovation market where three or more independently controlled research efforts in addition to those of the collaboration possess the required specialized assets or characteristics and the incentive to engage in R&D that is a close substitute for the R&D activity of the collaboration."[55] Although these "safety zones" do not apply specifically to mergers, similar consideration should apply for innovation effects from project-to-project mergers and mergers that combine overlapping R&D capabilities.[56] It is less clear whether such safe harbors should apply to product-to-project mergers because these types of mergers can have adverse effects for R&D effort and future price competition in a wide range of circumstances.

## 5   What Makes a "Good" Innovation Case?

Innovation cases range from relatively straightforward potential competition cases, in which a dominant firm acquires one of a very few entities that have the ability and incentive to disrupt the industry, to cases that involve many projects at an early stage of R&D with long lead times and uncertain future market impacts. This section summarizes characteristics that generally support antitrust intervention for mergers that may threaten innovation and future price competition. Agency experience with merger cases that allege harm to innovation informs this taxonomy. For the reasons explained previously, evidentiary hurdles are lower for product-to-project mergers than for project-project mergers and they are higher still for mergers of firms with overlapping R&D capabilities. Within these broad categories, other factors are relevant to the strength of an antitrust enforcement case that alleges harm from a merger for innovation or future price competition.

Chapter 7 describes some notable merger cases with innovation allegations and their lessons for the design of effective remedies. Some of these cases, such as the proposed merger of the mechanical heart pump manufacturers Thoratec and HeartWare, score highly on these characteristics. Other cases discussed in chapter 7 lack some of these characteristics and hence are more challenging examples to establish likely harm to innovation or future price competition.

## The relevant research and development market
## is highly concentrated
The concept of the research and development market is a tool to assess concentration in R&D effort directed to similar ends. Participants in the R&D market are the firms that possess the specialized assets necessary to innovate in the same product space. Concentration in the market can be measured by R&D expenditures, citation-weighted patents, or other indicators of innovative effort and capacity.

There is no simple formula to determine a bright line above which concentration in R&D warrants antitrust intervention for a merger. Downward innovation pressure is a unilateral effect from any merger, but its magnitude is small if diversion between the merging parties is a small fraction of diversion between each of the merging parties and the universe of rival innovators. That is likely to be the case if R&D effort is not concentrated. Furthermore, rival responses can offset unilateral downward innovation pressure, as can R&D efficiencies and the benefits from intrafirm technological spillovers. These qualitative arguments suggest that, consistent with the safety zones in the IP guidelines, a merger is unlikely to cause a significant reduction in R&D effort if, postmerger, five or more firms possess the necessary assets and have incentives to compete with the merged firm in R&D directed to similar ends. That presumption can be reversed, if as discussed next, there are significant disincentives for innovation from replacement effects that are created or enhanced by the merger.

### The merger creates or enhances replacement effects

Innovation incentives can be severely attenuated if firms have profits that are at risk of being replaced by an innovation. A merger would increase replacement effects if it creates or enhances profits that would be at risk from successful innovation by either merger partner. That could be the case even if the merging parties' existing products are not substitutes for each other, provided that profits from these products are put at risk from an innovation. If the merger increases prices and profits for the parties' existing products because they are substitutes for each other, that is an additional reason why it can increase the replacement effect.

### Innovations are new products for which appropriation is high

Appropriation is a catch-all term that reflects the ability of an innovator to capture value from the innovation. If appropriation is high, a firm can profit from a new product by taking business from rivals and by expanding sales.[57] All else being equal, a merger reduces the incentive to innovate because each merging party has one less rival from which it can steal business. The downward pressure from a merger for R&D effort is larger if successful innovation would allow a merger partner to appropriate a large fraction of its partner's business if they did not merge, and if that business is highly profitable. On the other hand, a merger can incentivize innovation if it increases the ability of the merging parties to profit from an innovation.

Strong intellectual property rights facilitate appropriation, but they are neither necessary nor sufficient to benefit from discoveries. Strong

IP rights are not necessary because firms often have other ways to appropriate benefits from an innovation, including lead-time and specialized assets that enable the innovator to create value and are not available to rivals. Conversely, strong intellectual property rights do not guarantee that inventors will profit from innovations. Consider the market for statin drugs that help to lower cholesterol levels in the blood. There are numerous statin variants, including atorvastatin, fluvastatin, pravastatin, rosuvastatin, simvastatin, and pitavastatin, all of which have or have had patent protection. Patents did not prevent competition that reduces profits from new statin discoveries.

### Intrafirm technological spillovers are low

Intrafirm technological spillover measures the extent to which a discovery by one merger party benefits the other party. If there are no intrafirm spillovers, the parties to a merger are essentially independent innovators with centralized control of pricing and output decisions. Under these conditions, if there are no other merger-specific benefits, the merger creates unilateral downward innovation pressure by internalizing the business-stealing effect that would have occurred if the merging firms remained as independent competitors.

In contrast, large intrafirm spillovers can increase postmerger innovation incentives by allowing each party to a merger to share innovation benefits that accrue to the other party. The example of a cost-reducing innovation discussed earlier in this chapter illustrates the importance of intrafirm spillovers. If a new technology lowers production costs and if the technology cannot be profitably licensed to others, the benefit from the new technology is proportional to the innovator's sales. A merger expands the sales that can benefit from the new technology if intrafirm spillovers are large, provided that the merged firm does not raise prices to an extent that overwhelms the benefits from intrafirm technological spillovers. When intrafirm spillovers allow the merged firm to appropriate greater value from innovations, they can increase innovation incentives compared to the premerger incentives for each firm to invest to lower its production costs. A similar argument applies to product innovations.

Careful analysis of merger effects on innovation requires a balanced approach to assessing possible harms and benefits. To further that objective, the antitrust agencies should pay particular attention to the potential for intrafirm spillovers that would enhance postmerger innovation incentives.

**There is the expectation of large future price effects**

One category of potential adverse effects from a merger is harm to competition in markets that do not yet exist but would be created by the R&D efforts of the merging parties. Future price effects in these new markets are a separate concern from the price effects for existing products or harm to innovation.

Some mergers are more likely than others to have predictable price effects in new markets. These effects would be particularly difficult to identify and assess if innovations relate to broad categories of products for which the constellation of likely competitors cannot be established with any certainty. For example, a merger of two of only a few firms engaged in oncology R&D could have predictable effects on the development of new cancer drugs, but it would be difficult to assess the future price effects without knowing more about the drugs' clinical targets, therapeutic benefits, and available treatment alternatives.[58] In contrast, it is likely that a merger of two of only a few firms engaged in R&D for automatic transmissions used in large trucks would adversely affect prices and innovation for new or improved transmissions for these vehicles.

**The merging parties are likely innovators in the near term**

As discussed previously, the mere fact that R&D has a low probability of success does not prevent harm from a merger or justify a decision not to challenge the merger. For similar reasons, the time required to realize a successful innovation should not be the determinant factor in an enforcement decision related to harm for innovation or future price competition from a merger. However, innovations and future innovators are hard to predict, and antitrust agencies should factor uncertainty into their enforcement actions. Time to fruition is a relevant factor because future developments can make current R&D efforts obsolete or create substitutes that lessen the impacts of any reduction in effort from a merger. All else being equal, the difficulty of predicting competitive effects for innovation and future price competition increases with the time required to complete R&D projects and to realize their benefits as commercial products.

**The merging parties do not have strong innovation incentives that are independent of competition**

Parties can have powerful incentives to innovate that are largely independent of competition. As noted in chapter 4, the manufacturer of a durable good can profit from sales to new customers and to existing customers that choose to upgrade to a new version of the durable good. One way to increase sales is to innovate to attract more upgrades. A merger may have little adverse effect on innovation incentives for durable goods that are driven by upgrades compared to an otherwise similar merger of producers of nondurable goods.

### Acquisition is not necessary to incentivize innovation

Acquisitions of promising start-ups by dominant firms risk eliminating potential competition that could disrupt established monopolies. Antitrust authorities should be on alert to prevent these types of acquisitions.

There is, however, an important trade-off. Some firms are motivated to invest in R&D by the prospect of a buy-out. Venture capitalists invest in many high-tech start-ups with the expectation that, if successful, they will be sold to established companies. The pharmaceutical industry alone witnessed more than 1,200 mergers and acquisitions in the years 2014–2016, totaling more than $750 billion in aggregate total deal value.[59] Some of these acquisitions may have eliminated potential rivals. But other acquisitions rewarded innovations that would not have occurred if the entrepreneurs could not sell their R&D assets or license their discoveries to established companies. Many of these acquisitions combine complementary assets, such as R&D, clinical testing, marketing, and distribution, that cannot be economically duplicated by either the acquiring or the acquired firm.

A prohibition on acquisitions would discourage innovation, and consumers would be worse off if the number of discouraged discoveries exceeded the number of products suppressed by acquiring firms. Antitrust authorities can take a middle road: They can assess the extent to which acquisitions by established firms motivate innovations and explore other merger alternatives. This proposal does not require that courts abandon their traditional function to evaluate transactions solely on their merits and instead become industrial planners. Rather, it suggests that courts should put less weight on claimed benefits from a merger when there are alternative acquirers that have incentives to acquire a target and would have much less risk of harm to innovation or future price competition. These less-restrictive alternative acquisitions can preserve innovation incentives for entrepreneurs seeking buyouts without threatening innovation or future price competition.[60]

# 6  Competition and Innovation: Empirical Evidence

The most important common feature of the few R&D and innovation analyses that have sought to control for the underlying technological environment is a dramatic reduction in the observed impact of the Schumpeterian size and market power variables.
—William L. Baldwin and John T. Scott, "Market Structure and Technological Change" (1987)

## 1  Introduction

Researchers have attempted to identify and measure a causal relationship between firm and industry characteristics and the pace of innovation for decades. Wesley Cohen and others provide comprehensive surveys of the historical record.[1] Rather than repeat these surveys, I focus in this chapter on several recent empirical studies that reflect the current state of the art for the econometric analysis of the relationship between competition and innovation and highlight studies that are relevant to mergers, a key antitrust policy lever.

The economic incentive to innovate depends on technological opportunities and the nature of the innovations as well as on industry and firm characteristics. Section 2 describes the many factors that complicate attempts to determine a causal relationship between competition and innovation. Section 3 summarizes the results from several studies that compare the effects of competition in different industries on investment in research and development (R&D) or the output of innovations and best attempt to control for these confounding factors.

Section 4 examines empirical studies of R&D competition in selected industries with a focus on markets for personal computer (PC) hard disk drives (HDDs) and microprocessors. The microprocessor study highlights the influence of product durability on investment incentives and shows that even a monopolist has an incentive to innovate to sell upgrades to customers that have already purchased its product. This section also reviews empirical evidence related to whether incumbent firms tend to invest in product and process improvements or more radical innovations.

Section 5 surveys empirical studies of the effects of mergers on innovation. These studies are particularly relevant to the theme of this book because merger policy is the most common way that antitrust enforcers regulate competition. Furthermore, as noted in chapter 5, the effects of a merger on incentives to invest in R&D are not equivalent to the effects from a reduction in competition, which underscores the policy significance of empirical results from studies of actual mergers.

Section 6 offers some case studies of policy interventions that have had consequences for innovation and future price competition. These include the breakup of AT&T and several case studies and related research that examine the effects of compulsory licensing of intellectual property (IP) on innovation. Antitrust agencies have settled numerous merger challenges with requirements that IP owners license their intellectual property either without royalties or at reasonable terms. In the more distant past, courts and antitrust authorities have compelled dominant firms to license their intellectual property, often with positive consequences for industry innovation and price competition. The final section concludes with a brief summary of the empirical evidence relating to competition and innovation.

## 2   Unraveling the Industry-Innovation Connection

Early econometric studies found a relationship between industry concentration and innovation measures that displayed the shape of an inverted U.[2] The result was greeted with enthusiasm because it was consistent with both the teachings of the Arrow replacement effect, which implies that monopoly power discourages innovation, and Schumpeterian notions that highly competitive markets can be inimical to innovation by limiting the ability of innovators to appropriate value from their discoveries. However, these early studies failed to account for many factors that confound the relationship between competition and innovation. More recent studies that correct for these deficiencies demonstrate a more nuanced relationship between innovation and firm and industry characteristics.

### The measurement of innovation

Empirical studies have used R&D expenditures, counts of patents and innovations, and revenues as alternative measures for the output of innovation. R&D is a flawed measure because it is an input to innovation rather than a measure of the output of innovation. R&D expenditures can increase with no corresponding benefit for innovation, and efficiencies can allow firms to decrease their R&D spending without harming innovation. Patents are also imperfect measures of innovation, even when adjusted for quality, which is often done by using citations from other patent applications. Patent values are highly skewed, and in industries such as semiconductors, firms often patent to defend against the threat of costly litigation and to negotiate access to external technologies on more favorable terms. These factors make patent counts a weak indicator of the progress of technology in these industries.[3] Revenues reflect the importance of innovations but also include the confounding effects of market power.

The measurement of innovation is also complicated by the fact that it is often difficult to trace its source. Innovations often come from unexpected directions, including from firms in unrelated industries or individual inventors. An example is the semiconductor manufacturing industry, where photolithography technologies that enabled the miniaturization of semiconductor circuitry had their origins in a different industry, optical instruments.[4] For many industries, the sources of invention and its economic benefits are numerous, scattered, and varied.

### Innovation, firm size, and competition are codetermined

Studies that attempt to identify a causal relationship between firm and industry characteristics and the output of innovations must contend with the circular relationship between innovation, firm size, and competition. They are codetermined. Larger firms may invest more in R&D, but size is also a consequence of past success. Innovation confounds measures of competition because market dominance can reflect successful innovation by a firm with better products or superior production technologies. It is also possible that innovation can open the gate to new entry into an industry or expansion of smaller incumbents, leading to a decrease in market concentration and an increase in apparent rivalry. Successful innovation by a firm that is far from the technological frontier can create new competition by narrowing the gap in quality or production costs relative to the market leader, but empirical studies may fail to account for changes in competitive pressure from firms' relative technological capabilities.[5]

In the language of econometrics, firm size, market concentration, and market power are endogenous to the variable—innovation—that the econometrician is attempting to explain. Early studies made little or no attempt to control for this endogeneity, which led to erroneous conclusions about the causal relationship between firm and industry characteristics and innovation. More recent studies recognize the problem and use various techniques to attempt to control for these interdependent effects.

### Controlling for technological opportunity and appropriation

The relationship between industry and firm characteristics and the output of innovations depends on the available technological opportunities, which also tend to be correlated with firm and industry characteristics. The significance of competition or industry concentration as an independent determinant of R&D expenditures or innovation faded when researchers repeated early studies with variables that accounted for industry differences.[6]

Investments in R&D have economies of scale for many technological opportunities, which can cause observed investments to be correlated with firm size and market concentration.[7] An observation that scale or concentration is correlated with innovation may be nothing more than an observation that firms tend to be larger when R&D is a significant expense. That point, by itself, does not imply that firm size causally determines innovation.

Incentives for investment in R&D depend on the ability of a firm to profit from the value of inventions that may emerge from those investments, which are related to many factors, including patent rights, opportunities for trade secrecy, benefits from lead time, and complementary factors of production and marketing that can serve as barriers to competition from firms that may want to imitate an invention. These appropriation factors depend on the technology and the industry in which the firm resides and can change over time. In addition, they may differ significantly for process inventions that offer the opportunity to achieve lower production costs and for new products that can expand a firm's sales.[8]

### Data limitations

The available data to test these relationships is often lacking. Innovation is difficult to measure. Some researchers, such as the Science Policy

Research Unit (SPRU) in the UK, have produced databases of innovation counts, but the counts do not cover all innovations and only allow crude classifications of their significance.

Accounting for R&D expenditures is often imprecise, and R&D data is frequently absent for small firms and for private firms that are not required to report their expenditures. This can lead to comparative underreporting by smaller firms, which are disproportionally private. Firms often in-license technology rights, which may or may not be recorded as an R&D expense. Most firms comprise many business units that operate in different industries. The relevant technological opportunities and industry effects exist at the level of the business unit but, with a few exceptions, data is at the firm level. It is important to recognize that R&D expenditure is an input to innovation and is not a measure of the output of innovations.

A related data concern is the measurement of market power. Studies often use market concentration or firm share as an index of market power. Market concentration and firm share depend on the definition of the market, which often bears little relation to the relevant competition. Often data is only available at a level that aggregates over many distinct products, which undermines its value as an indicator of actual competition, or markets are defined too narrowly and consequently fail to account for important sources of competition. High market concentration or firm share does not necessarily indicate market power if consumers would easily switch suppliers in response to higher prices, and low concentration or firm share does not necessarily indicate a lack of market power if consumers are reluctant or unable to switch suppliers.

### Statistical limitations

Modern statistical methods often rely on differences between separate but otherwise similar populations to measure relevant effects. For example, studies of the effects of competition on market prices can exploit differences in competition across local geographies. That is difficult to do for innovation because innovation often has a global, or at least national, geographic dimension. Hence, empirical studies of competition and innovation have to resort to other, generally less satisfying, ways to control for the effects of confounding factors in the competition-innovation relationship. The difficulties of constructing valid tests of the relationships among innovation, firm characteristics, market structure, and competition warrant caution in interpreting empirical results despite numerous attempts to test these relationships.[9]

### Productivity versus innovation

"Productivity" is a measure of the efficiency of production. "Labor productivity" is a measure of output per worker. "Total factor productivity" accounts for capital and other inputs in addition to labor.[10] Many empirical studies show a positive relationship between competition and labor or total factor productivity.[11] Although merging parties often claim that the merger will yield efficiency benefits, there is little empirical support for a presumption that mergers generally increase productivity.[12]

It is tempting to infer from this body of empirical research that competition promotes innovation and, conversely, that mergers harm innovation. However, most productivity studies do not support this conclusion, because an increase in productivity is not the same as an increase in innovation. Competition can increase productivity in two ways. One mechanism is a Darwinian selection effect, which forces inefficient plants to close and creates incentives for firms to allocate resources to more productive activities. A second mechanism operates through process innovations that lower production costs or new products that increase output. Only the second mechanism reflects innovation by the firms that survive competition, but few productivity studies distinguish between these two effects.

Some studies demonstrate a positive relationship between competition and the rate of growth in productivity.[13] Productivity growth is also a flawed measure of innovation because it measures the historical rate of change in a static measure of the efficiency of production. An increase in productivity could be caused by increases in competition that are unrelated to innovations and that cause the growth of more efficient firms and the exit of less efficient firms over time through the Darwinian selection mechanism. The relevant empirical research for the competition-innovation nexus relates competition, market structure, or mergers to a measure of innovative output or a surrogate measure, such as investment in R&D or counts of patents or new products. I focus on these studies.

## 3   Interindustry Econometric Studies

With these caveats in mind, I briefly summarize a few studies that utilize modern econometric methods to assess the relationship between industry structure and innovation. Richard Blundell, Rachel Griffith, and John Van Reenen employ a sample of 340 manufacturing firms listed on the London International Stock Exchange over the period 1972–1982 to explore the relationship between firm size, market concentration, and innovation.[14] They measure innovation using innovation counts from the SPRU, along with patent data. The SPRU counts an innovation if it is technologically significant and commercialized by a firm. The authors attempt to control for the importance of heterogeneity in firms' innovation capabilities by calculating a variable that measures the stock of the firm's innovations prior to the start of the sample period.

Their study finds that more-concentrated industries produced fewer innovations. Within industries, firms with large market shares introduced more innovations, but there is no correlation between absolute firm size and the significance of innovations, and no evidence that cash flow is a significant determinant of innovation.[15] This result undermines the Schumpeterian view that market power provides a stable platform that promotes investment in R&D. The authors find similar results when they use patents as a measure of technological performance, and when they examine a subset of data confined to the pharmaceutical industry, which is notable for its high R&D intensity and for the importance of patent protection.

The empirical analysis by Blundell and his coauthors covers a short time frame that preceded the internet age, and it is limited to UK firms. Furthermore, the authors assume a linear relationship between market concentration and innovation. They do not investigate whether there is a nonmonotonic relationship such as the inverted-U relationship from early studies that failed to control for industry and firm characteristics. A 2005 study by Philippe Aghion and coauthors specifically address this latter issue.[16] They follow patents issued to 311 firms listed on the London Stock Exchange over the period 1973–1994, weighting each patent by the number of times that it is cited by another patent.

Aghion and his coauthors use an industry average of firm-specific Lerner Indices to measure competition. The Lerner Index measures the extent to which a firm exercises market power by raising its price above its marginal production cost.[17] In this respect, the average Lerner Index can be a better measure of industry competition than a concentration index, although that depends on the firms that are included in the average. An additional concern is that the calculation relies on accounting data that may differ significantly from a firm's actual marginal cost, which makes the Lerner Index a less reliable measure of market power.

Adding controls for industry-specific innovation propensities, the authors find that innovation output, as measured by citation-weighted patents, displays an inverted-U relationship with the industry average Lerner Index. Although these results are intriguing, they should be interpreted with caution because the index calculations are for broad industry classifications, and it is debatable whether the authors' industry-specific controls fully account for differences among firms in their ability to innovate or the fact that competition and innovation are mutually determined.

The authors address some of these issues by repeating the analysis using a set of policy measures that caused unanticipated changes in competition. These are industry privatizations that occurred in the Margaret Thatcher era, the European Single Market Programme that liberalized trade across member nations in 1988, and investigations by the UK Monopolies and Mergers Commission that resulted in structural or behavioral remedies. Applying these policy instruments as controls in their empirical analysis does not change the inverted-U relationship between patenting and the industry average Lerner Index, but the controls do not erase concerns about the aggregation of indices in broad industry classifications.

In a subsequent paper, Philippe Aghion and several coauthors drill down further to explore the effects of new competition on innovation.[18] They measure new competition by the entry of new production facilities by foreign firms and measure innovation by patent counts for 174 UK firms over the period 1987–1993.[19] They also keep track of the distance of each industry from the technological frontier, which they measure as the gap between the average labor productivity of the incumbent UK industry and the labor productivity of its counterpart in the US (because UK productivity generally lagged productivity in the US over this time period).

The authors find an inverted-U dependence of patenting on industry competition measured by average profitability, but the effects differ dramatically for industries close to and far from the technological frontier. New competition spurs patenting for industries close to the frontier but has the opposite effect for industries far from the frontier. The authors argue that these differential effects are consistent with their theoretical model of stepwise innovation incentives. Firms that are close to the frontier have an incentive to invest in R&D to escape head-to-head competition from efficient new entrants, while the entry of efficient new competitors discourages investment by firms that are far from the technological frontier by lowering the return from catching up to industry leaders.

Richard Bloom, Mirko Draca, and John Van Reenen study the effects of competition on innovation by European firms that manufacture clothing and textiles.[20] While not generally branded as high-tech, firms in these sectors held over 30,000 European patents in the ten-year period, 1996–2005, covered by the study. They focus on import competition from China's entry into the World Trade Organization in 2001 and measure innovation by citation-weighted patents and investments in R&D and information technology.[21] The study finds that surviving firms exposed to Chinese import competition filed more patents and increased their investments in R&D and information technology. The authors do not find any statistically significant effects on patenting or other innovation-related measures in response to increased import competition from developed countries. One explanation is that Chinese imports were a wake-up call for firms that had the ability to differentiate themselves technologically from low-wage competition. Competition among technologically advanced firms offered less scope for differentiation.

German Gutiérrez and Thomas Philippon examine investment in R&D, as well as other measures, for forty-three groupings of US nonfinancial firms in response to import competition over the period 1995–2015. They find a positive relationship between competition and R&D investment. Consistent with results in other studies, they find that industry leaders (defined as the firms that account for the largest one-third of market value) responded more to the shock of Chinese import competition than industry laggards (firms in the bottom third of market value).[22]

However, a study that examined patenting activity and R&D by US manufacturing firms over the period 1991–2007 reports the opposite conclusion. David Autor and coauthors document a negative relationship between US firms' exposure to Chinese competition and their patent applications and investments in R&D. The authors conclude that "the innovation response of U.S. firms more exposed to rising market competition from China has been substantially and unambiguously negative."[23] The authors' explanation for their results is that firms scaled back their global operations in response to shrinking demand caused by global competition, and that contraction included a reduction in R&D expenditures and patenting.

This study includes patenting by textile manufacturers, yet reaches opposite conclusions from those reported by Bloom, Draca, and Van Reenen for European firms. Autor and his coauthors reconcile their contrasting findings by showing that the negative effects of import competition concentrate on firms that are initially less profitable and capital intensive. This is consistent with theoretical results about the relationship between competition and innovation and with the findings in other empirical studies, which show differential effects of competition on innovation by firms with different characteristics.[24]

Table 6.1 summarizes the conclusions from these interindustry studies for the effects of competition and industry structure on innovation. Unfortunately, these studies do not reach a consensus, other than to note that innovation effects can differ dramatically for firms that are at different levels of technological sophistication. Although some studies find a positive relationship between measures of innovation and competition (alternatively, a negative relationship between innovation and industry concentration), others find that the relationship exhibits an inverted-U, with the largest effects at moderate levels of industry concentration or competition, and at least one study reports a negative relationship between competition (measured by Chinese import penetration) and innovation (measured by citation-weighted patents and R&D investment). One consistent finding is that an increase in competition has less of a beneficial effect, and may have a negative effect, on innovation incentives for firms that are far behind the industry technological frontier.

The inconsistent findings from these studies reflect the difficulty of measuring competition and innovation and unraveling their codependence. Moreover, these studies have only limited relevance for competition policy. The sources of new competition in several of these studies arise from trade policies, which promote foreign direct investment and have possible additional effects by expanding markets for exporters, crowding out demand for domestic suppliers, and lowering the prices of intermediate inputs. Competition policies, such as merger enforcement, do not generally have similar trade-enhancing consequences.

Studies that focus on a particular industry can help to isolate determinants of the competition-innovation relationship that are specific to industry capabilities and technological opportunities. The next section reviews some of these studies. Section 5 describes studies that address the effects of mergers and acquisitions on measures of R&D investment and innovation. These studies have particular relevance to antitrust

**Table 6.1**
Summary of interindustry studies of competition and innovation.

| Study | Firms and Time Frame | Innovation Measure | Competition Measure | Conclusions |
|---|---|---|---|---|
| Blundell et al. (1999) | 340 UK manufacturing firms; 1972–1982 | Innovation and patent counts | Industry concentration | Competition and firm share *positively* related to innovation and patenting |
| Aghion et al. (2005) | 311 UK firms; 1973–1994 | Citation-weighted patents | Lerner Index | *Inverted-U* relationship between competition and patenting |
| Aghion et al. (2009) | 174 UK firms; 1987–1993 | US registered patents | New entry by foreign firms | *Positive* relationship between competition and patenting for firms close to the technological frontier. *Negative* relationship for firms far from the technological frontier. |
| Bloom et al. (2016) | European manufacturers of clothing and textiles; 1996–2005 | Citation-weighted patents, R&D and information technology (IT) investment | China entry into World Trade Organization; other entry | Competition *positively* related to patent counts and investment in R&D and IT. *No response* to entry from developed countries. |
| Gutiérrez and Philippon (2017) | 43 groupings of US nonfinancial firms; 1995–2015 | R&D investment | Industry concentration; China import penetration | *Positive* relationship between competition and patenting for industry leaders. *Negative* relationship for laggards. |
| Autor et al. (2020) | US firms in almost 400 industries; 1991–2007 | Citation-weighted patents and R&D investment | Chinese import penetration | Competition *negatively* related to patenting and R&D investment. Larger negative impact for weaker firms. |

policy because the decision to block a merger or condition approval on a behavioral or structural remedy is the most common tool used by enforcement agencies to address innovation effects.

## 4   Industry Studies

Empirical studies that mine data from many industries typically attempt to control for differences among broad industry classifications, but that can fail to capture differences in technological opportunities that exist at more of a microlevel. Single-industry studies avoid some of this variation, although technological opportunities within a single industry can and do change over time. Nonetheless, single-industry studies can provide a fertile field to test theoretical predictions about market structure and innovation.

### R&D "races"

One candidate for testing is the relationship between market structure and patenting in industries that have characteristics of competition in which the winner takes all (or at least most) of the value from a discovery. An example is a race to patent the next important invention. Jennifer Reinganum argues that incumbents have less incentive to invest in R&D than new entrants because innovations replace their existing profits.[25] This Arrow replacement effect implies that incumbents are less likely than new entrants to patent the next important discovery. In contrast, Richard Gilbert and David Newbery show that a dominant firm has an incentive to invest to preserve its dominance and, in some circumstances, is more likely than an entrant to make the next discovery.[26] Clayton Christiansen claims that incumbents fail to sustain technological leadership because they focus too much on the needs of their immediate customers and ignore emerging technologies that ultimately displace established practices.[27]

Economists have applied econometric analysis to unravel these contrasting arguments. Josh Lerner examines competition between HDD manufacturers and the characteristics of firms that developed higher-density drives over nearly two decades. He concludes that firms in the HDD industry that lagged the current technological leader—and particularly those in the middle of the technological pack—had a greater propensity to innovate than the leader and were more likely to win the race for the next generation of disk drives. His results are consistent with both the replacement effect and Christiansen's theory of incumbent customer focus.[28]

Mitsuru Igami develops a dynamic model of HDD investment and applies the model to the transition from 5.25-inch to 3.5-inch HDDs, which were common formats in PCs over the relevant period of his analysis. Igami's model allows him to sort out three contrasting explanations for innovation by incumbents and new entrants: (1) the Arrow replacement effect, (2) preemption, and (3) differential investment costs. Igami finds some evidence to support preemptive investment by incumbents. Nonetheless, he confirms previous results that find that incumbents invested less in the new 3.5-inch format than new entrants. The dynamic model attributes this result to the Arrow replacement effect; incumbents were reluctant to invest in a new product that would cannibalize their existing products. The replacement effect is strong enough to more than offset his empirical conclusion that incumbents have a cost advantage for new product investments relative to entrants.

These results illustrate the relative importance of replacement effects and preemption incentives for innovation, but they are limited to the HDD industry; other industries need not evidence similar behavior.[29] Furthermore, the HDD industry is not a good candidate to test preemption incentives. More than twenty firms sold 5.25-inch HDDs when 3.5-inch HDDs first arrived in 1982 (although four to six firms accounted for more than 50 percent of sales by the late 1980s). Preemption incentives are weak in an oligopoly because a firm that engages in costly preemptive investment has to share the benefits of preemption with its rivals. It is significant that Igami identifies a role for preemption in the HDD industry, notwithstanding the relatively unconcentrated structure of the industry for most of the period spanned by the Igami study.

## Durable goods

Many innovations create new or improved durable goods. A firm that sells (rather than rents)[30] a durable good has three ways to profit from additional sales: it can benefit from sales to new consumers that were not previously in the market; it can lower its price to induce purchases by existing consumers that did not want to buy the good at the higher price; or it can offer a new and improved product to attract new consumers and induce existing customers to upgrade their purchases. The first alternative is not available in a mature market with no consumer growth. The second alternative can result in an unprofitable downward price spiral if the firm cuts prices to attract more demand. That leaves innovation as an appealing strategy for producers of durable goods, even for firms that command a large share of the market.

As noted in chapter 3, the Arrow replacement effect does not have the same deterrence effect for a monopoly producer of a durable good because even a monopoly seller of a durable good has an incentive to innovate to increase its sales to consumers who choose to upgrade their purchases. Unfortunately, little empirical research has addressed the relationship between competition and innovation specifically for durable goods.

An exception is the study by Ronald Goettler and Brett Gordon of competition and innovation in the microprocessor industry. The microprocessor industry is an excellent candidate to explore the effects of competition and antitrust policy on innovation for durable goods. Microprocessors are durable goods; they can function without depreciation for many years. A consumer's incentive to replace a microprocessor is often the desire to upgrade to a more powerful processor. The industry has been a duopoly for decades, with Intel and Advanced Micro Devices (AMD) accounting for about 95 percent of desktop PC microprocessor sales. Innovation can be measured by the clock speed of the microprocessor. Both Intel and AMD have invested heavily in microprocessor R&D. Clock speeds for the newest microprocessors roughly doubled every seven quarters over the twelve-year period examined by Goettler and Gordon (1993–2004).

Intel has been the target of several antitrust actions related to conduct that allegedly excluded AMD from microprocessor sales. In 2009, Intel paid AMD $1.25 billion to settle charges that Intel violated antitrust laws by offering rebates to Japanese PC manufacturers who agreed to eliminate or limit purchases of rival microprocessors.[31] In the same year, the European Commission (EC) fined Intel 1.06 billion euros and ordered the company to end its rebate program.[32] In 2010, the US Federal Trade Commission (FTC) and Intel settled charges that Intel unlawfully maintained a monopoly in microprocessors and attempted to acquire a second monopoly in graphics processors using a variety of unfair methods of competition.[33]

Goettler and Gordon propose a dynamic model of microprocessor investment. After estimating the model parameters using data from 1993–2004, the authors pose counterfactuals to explore how different market structures would affect their conclusions regarding industry structure, R&D investment, and pricing.

One of their counterfactuals compares R&D investment and pricing in the actual market to investment and pricing for a hypothetical Intel monopolist. They reach the striking conclusion that microprocessor innovation would have been *greater* if competition from AMD had been eliminated over the study period. The reasons for this counterintuitive result can be traced to the demand for upgrades and investment incentives from higher prices. Most of the demand for PC microprocessors came from upgrades over the study period; for instance, 82 percent of PC sales in 2004 were replacements of existing units.

Intel's microprocessor prices would have been higher without competition from AMD. The Schumpeterian appropriation effect from higher prices increases the incentive to innovate, but this innovation benefit comes at a large consumer cost. The authors conclude that higher prices from a hypothetical Intel monopolist would more than offset the benefit from greater innovation; consumers would be worse off. This result repeats findings from theoretical models discussed in previous chapters. Higher prices can promote innovation, but often the higher prices imply that consumers fail to benefit from the innovations that they foster.

### Incremental versus drastic innovation

Concerns about the effects of rising market concentration in the US economy on innovation are not limited to the effects of concentration on the level of R&D investment. An additional concern is that large firms in concentrated industries have a bias toward incremental improvements in existing products and technologies rather than transformative innovations.[34]

There are theoretical justifications for concern about the effects of firm size and market concentration on the direction of innovation. Chapter 3 explains why firms in concentrated industries can have incentives to invest in R&D to preempt rivals, but those incentives are absent if innovation is drastic. Furthermore, established firms can face large disruption costs for radical innovations that require the firms to change the internal architecture that defines their ways of doing business. This organizational effect, emphasized by Rebecca Henderson[35] and others, can bias large established firms in favor of incremental innovations.

Wesley Cohen surveys empirical studies that conclude that larger firms pursued more incremental innovations and more innovations related to process improvements.[36] The positive relationship between size and process innovation has a clear theoretical explanation because larger firms have more to gain by lowering their production costs. In a more recent study, Daniel Garcia-Macia, Chang-Tai Hsieh, and Peter Klenow use employment data to make inferences about the types of innovations pursued by incumbents and new entrants.[37] They equate large changes in employment to disruptive innovation (the "creative destruction" described by Schumpeter) and smaller changes to incremental innovation. With the qualification that employment changes can be explained by factors other than innovation, the authors conclude that incumbents tend to make incremental quality improvements rather than develop disruptive new products. Nonetheless, most growth comes from incumbents rather than entrants, primarily because entrants account for a small fraction of industry employment and innovation.

## 5  Mergers

Merger enforcement is the single most common policy lever employed by competition authorities, and nearly all merger challenges in high-tech industries include an allegation of harm to innovation.[38] Yet empirical evidence for the effect of mergers on innovation is sparse. The empirical literature on the relationship between competition and innovation, summarized in table 6.1, provides uncomfortably weak support for the conclusion that increased competition in concentrated markets promotes innovation for technologically advanced firms; however, a merger is not the same as a reduction in competition.

In addition to the factors that confound empirical studies of competition and innovation, a complication for empirical studies of mergers is that they take place in the shadow of antitrust enforcement. The Hart-Scott-Rodino (HSR) Act requires companies to notify the antitrust authorities about a planned merger or acquisition if the transaction exceeds modest thresholds, and the authorities will block a merger if they believe that it is likely to raise prices or harm innovation or condition the merger on remedies that are intended to eliminate anticompetitive effects. If they could do their job perfectly, no mergers would cause consumer harm. That is not possible, though, because merger enforcement is a prediction that is subject to error, and the antitrust authorities often miss mergers that fall below the HSR reporting thresholds.[39]

There are errors of underenforcement and overenforcement. Merely observing that *some* mergers harmed innovation is not enough to conclude that the agencies should be tougher on *all* proposed mergers.

Some bad mergers will escape challenge by antitrust authorities, and some good mergers will be blocked. Furthermore, the mere observation that merged firms invested less in R&D than firms that did not merge is not sufficient to conclude that mergers harm innovation. Firms that choose to merge are typically different from firms that do not merge. Firms may have an incentive to merge because, acting independently, they anticipate poor performance from their R&D activities and expect that a merger partner will improve their prospects. As discussed next, this is a critical issue for the evaluation of mergers in the pharmaceutical industries, for which the breadth of merging parties' R&D pipelines and their complementary clinical and marketing capabilities can be important determinants of merger decisions. Consequently, merely comparing the performance of merged firms against a sample of firms that did not merge is likely to introduce significant error. Moreover, some mergers may lower R&D expenditures by eliminating redundancies without harming the output of innovations.

With these and other caveats in mind, empirical research has followed different tracks to estimate the effects of mergers on innovation or related measures, such as R&D spending or patenting. These include interindustry studies of merger effects, dynamic models that predict innovation outcomes, and studies of merger activities in single industries, such as pharmaceuticals, that have experienced large merger waves.

In their study of merger activity across many industries, German Gutiérrez and Thomas Philippon report a negative relationship between merger activity and investment.[40] The authors assume that large merger waves are mostly exogenous events and use the occurrence of these discrete waves to estimate effects on investment. The study shows that, conditional on measures of current concentration and expected sales growth, industries with greater merger activity are associated with relatively lower investment. This result should be viewed with caution because the study does not eliminate the possibility that merger waves are the result of declining expectations of R&D performance rather than the cause of a reduction in R&D investment.

Respondents to a survey of mergers conducted by Bruno Cassiman and coauthors reported that they increased R&D expenditures when they had complementary technological capabilities and reduced R&D expenditures if they operated in similar markets.[41] Robust conclusions from this undertaking are elusive because the survey is not a random sample, includes only thirty-one deals, and relies on self-reported descriptions of deal outcomes. While many of the respondents reported that the transactions had positive outcomes for innovation, the survey includes no calibration mechanism to compare innovation-related outcomes against likely outcomes for the parties if they had not merged. Without such a comparison, a positive report for innovation could merely reflect general postmerger satisfaction rather than an actual innovation benefit from the transaction.

Within-industry studies of mergers and acquisitions have followed one of two approaches. One approach is to estimate a dynamic model of pricing and R&D investment and then use the estimated parameters to assess the consequences from hypothetical mergers. Goettler and Gordon use a variant of this approach to explore the consequences of different market structures in the PC microprocessor industry. Mitsuru Igami and Kosuke Uetake also follow this approach to investigate the effects of hypothetical mergers in the market for PC HDDs,[42] and conclude that mergers in the HDD industry would have had only modest effects when they occur with more than three firms in the industry. Of course, these predictions are only as reliable as the models used to generate them.

A second approach exploits data from actual mergers in an industry that has experienced many mergers. As noted previously, this approach suffers from truncation bias—the data is unlikely to include the most troubling mergers because antitrust authorities would likely prevent or modify them.

Isolating the effects of mergers on innovation requires a necessarily imprecise method to predict the output of innovation postmerger and to compare that prediction with the output of innovation that would be expected if the firms do not merge. A firm may anticipate having excess capacity to manufacture and market drugs because its current drugs are nearing the end of their patent exclusivity and it expects to lose sales to generic equivalents. Another firm may have promising molecular entities in its pipelines with little capacity in place to manufacture and market the new drugs that may emerge from its R&D. A merger of the firms would allow better utilization of complementary factors of production, marketing, and R&D. Failure to control for these firm-specific attributes can lead to an erroneous conclusion that mergers harm innovation when the correct interpretation is that mergers are responses to declining innovation expectations by one or both of the merging firms.

As for other studies, it is important to distinguish the output of innovation from the input of R&D expenditures. Although an increase in postmerger R&D suggests an increase in expected innovation (if the increase is not wasteful duplication), the opposite is not necessarily true because mergers can increase the output of innovation while rationalizing redundant R&D expenditures.

Data limitations plague these studies. They may have data on acquiring firms, but only limited data on acquired firms. Some firms engage in acquisition sprees that make it difficult to identify the consequences from individual transactions. Few studies observe merger outcomes over time horizons that are long enough to fully identify innovation effects, and those that do have to control for additional factors that are unrelated to the merger and confound their conclusions. Another limitation is that most evaluations of mergers only examine innovation-related effects on the parties. A merger can change competitive conditions in an industry, which can increase or decrease innovation incentives for nonmerging firms. These R&D responses by nonmerging firms can offset or magnify effects observed for the merging firms.

Subject to these limitations, the pharmaceutical industry is a good candidate to study the effects of mergers and acquisitions on innovation performance for several reasons. First, mergers and acquisitions have transformed the pharmaceutical and biotech industries. Figure 6.1 shows a selection of major pharmaceutical deals that occurred during

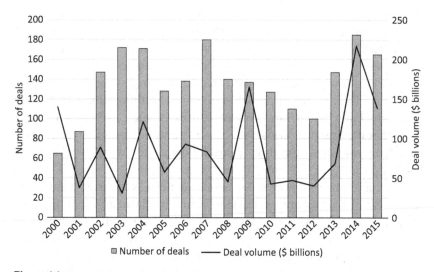

**Figure 6.1**
Selected merger and acquisition (M&A) activity in the pharmaceutical and biotech industries. Source: Visnji (2019).

the period 2000–2015. Second, the pharmaceutical and related biotech industries are among the most R&D-intensive sectors of the economy. Third, compared to some other industries, it is easier to measure the output of pharmaceutical innovations by counting either new drug applications or patents.[43] Fourth, compared to some other industries, it is easier to identify firm-specific innovation capabilities by examining projects in firms' clinical pipelines, for which there is public data.

Pharmaceutical R&D expenditures increased more than threefold over the period 1995–2015, while the production of new drugs hovered at around thirty per year over this interval. These statistics, in combination with the large measure of merger activity, suggest that pharmaceutical mergers have lowered R&D productivity; however, that confounds merger effects with a general industry trend of declining discoveries per R&D dollar as new drugs become more challenging to discover. The cost of developing a commercially successful new drug (including the opportunity costs of failures) increased from $271 million in 1987 to $2.6 billion in 2013 (unadjusted for inflation).[44] Other industries have also experienced large increases in the cost of R&D.[45]

A few empirical studies examine the effects of pharmaceutical mergers on drug discoveries and control for firm characteristics that might differentiate the innovation performance of firms that chose to merge from that of other firms in the industry. Carmine Ornaghi examines the performance of twenty-seven large pharmaceutical mergers and acquisitions over the period 1988–2004. To account for firm-specific differences between merging and nonmerging firms, the author creates a comparison set of firms using the merger propensity method. This approach estimates the likelihood that firms merge based on the percentage of their drugs approaching patent expiration, the percentage of newly launched drugs, and other observable characteristics. He finds that the merging firms did less R&D and produced fewer important drug patents in the three years following the merger, relative to his comparison set of firms.

Patricia Danzon, Andrew Epstein, and Sean Nicholson conduct a similar study that examines 165 transforming pharmaceutical mergers over the period 1988–2000. They define a transforming merger as a transaction valued at more than $500 million or representing more than 20 percent of a firm's premerger enterprise value. The study assigns firms to subpopulations of large and small firms. The large firms had an enterprise value of at least $1 billion in at least one year during the study period. Danzon and the coauthors also use the propensity method to match merging firms to a population of companies that had a similar likelihood of merging based on the number of the firm's marketed drugs, the fraction of drugs that are nearing the end of exclusivity, and other observable characteristics.

The study tracks the growth in enterprise value, sales, employees, and R&D expenses in the three years following mergers. The analysis demonstrates the importance of controlling for the characteristics of firms that are likely to choose to merge. Compared to the population as a whole, firms with a relatively high likelihood of merging in a particular year experienced relatively small growth in sales, employees, and R&D over the next three years regardless of whether they actually merged or not. This result shows that failure to control for conditions that lead firms to merge can result in overestimates of the negative effects of mergers on R&D investment and innovation.

Controlling for merger propensity, the authors report that the performance of large firms that merged was not significantly different from the performance of nonmerging firms. Small firms experienced some reduction in R&D in the first year following a merger, but the study also finds some beneficial effects from mergers for small firms that were likely to otherwise experience a reduction in performance due to adverse events such as the loss of drug exclusivity.

Justus Haucap, Alexander Rasch, and Joel Stiebale also conduct an empirical study of pharmaceutical mergers.[46] They examine sixty-five mergers reviewed by the EC between 1991 and 2007 and use patent applications as their measure of innovation. To estimate merger propensity scores, the authors use patent applications lagged one to three years before merger events, the presample average of the number of patents, the stock of citation-weighted patents, a variable indicating nonzero innovative activity prior to merger, sales, and the ratio of profits to sales. Unlike Ornaghi (2009) and Danzon, Epstein, and Nicholson (2007), this study does not include the percentage of a firm's drugs approaching patent expiration and the percentage of newly launched drugs in the calculation of merger propensity scores, which is a possible source of estimation error.

Haucap and his coauthors find a substantial decline in patenting postmerger. They estimate that three years after the merger, merging firms' patent stocks were more than 30 percent less than what they would have been if they had not merged. Furthermore, the authors report a knock-on effect from mergers: Averaging over their sample of mergers, the patent stocks of the merging firms' competitors were more than 15 percent less than what they would have predicted without the merger. Competitors followed reductions in patenting by the merging parties with reductions in their own levels of patenting. If the competitor responses that they observe hold more generally, this suggests that adverse effects on innovation by the merging parties underestimate the industrywide harm from a merger.

The results from these three studies are not consistent. Whereas Ornaghi and Haucap and his coauthors report significant declines in innovation-related activities following mergers, the study by Danzon and her coauthors finds no effect following mergers of large pharmaceutical firms. One explanation for the different conclusions is that the merger propensity method does not allow an exact comparison between merging and nonmerging firms in an industry, and the studies calculated merger propensity scores in different ways. Many relevant characteristics of firms are omitted or are not observable by an econometrician. The fact that some firms chose to merge and others did not, despite having similar merger propensity scores, suggests that the analysis overlooks key attributes that may be correlated with mergers and innovative performance. Nonetheless, these studies do not support a conclusion that mergers have promoted innovation in the pharmaceutical industry. This is an important empirical result, particularly in light of the fact that the observed transactions are those that did not raise sufficient competition concerns to be challenged by the antitrust authorities.

Henry Grabowski and Margaret Kyle follow a different approach to studying merger innovation effects by examining the progress of drug development projects following merger events. The authors utilize a database of more than 4,500 firms that were engaged in pharmaceutical R&D between 1990 and 2007. Their measure of R&D performance is the probability that a project in the R&D pipeline advances to the next stage of development (e.g., the probability that a project moves from phase II to phase III clinical trials). They find that for firms that experienced a merger during the 1985–2006 period, a higher fraction of their projects progressed to the next phase compared with nonmerging firms. The most substantial differences were in movements from phase III to the launch of a new drug. They observed this general trend for all size categories of firms.

The observations by Grabowski and Kyle do not indicate that mergers increase the output of pharmaceutical R&D. Their results could reflect greater commitment by merging firms to selected R&D projects, not necessarily greater allocation of R&D effort. The merged firms may have been better at advancing drugs through clinical trials because they abandoned less promising early-stage drug projects, which would indicate a decline in R&D effort. Furthermore, their analysis does not control for factors that lead pharmaceutical companies to pursue mergers.

A study by Colleen Cunningham, Florian Ederer, and Song Ma raises significant concerns about the effects of acquisitions in the pharmaceutical industry on the acquiring firms' allocation of R&D effort. In their aptly titled study, "Killer Acquisitions," they find that pharmaceutical companies are more likely to terminate drug research projects that they acquire from other companies compared to projects they originate themselves. The authors collected data on more than 60,000 drug projects. By 2017, companies had discontinued 85 percent of drug R&D projects that they had originated themselves in the period 1990–2011. This result reflects the well-known fact that most R&D projects fail to advance to commercial drugs. Over the same time period, they terminated 92.1 percent of R&D projects that they had acquired from other companies. Expressed differently, the companies advanced only about half as many R&D projects that they had acquired from others, compared to their internally generated R&D projects. The difference is statistically significant and implies that dozens of acquired pharmaceutical projects failed to become commercial drugs, compared to a hypothetical in which the projects have the same success rate as internally generated projects.

The authors conclude that technology and product overlaps explain most of the difference in termination rates for acquired pharmaceutical R&D projects. Technology overlaps are projects in the acquiring firm's R&D pipeline that have the same biological interaction as the acquired project. Product overlaps are acquired projects that target the same therapeutic class as a drug supplied by the acquiring firm. Product overlaps had a significant negative effect on the probability that the acquired firm continued to invest in the acquired project only when the product class was highly concentrated.

These results accord with the economic theory of R&D incentives. Product overlaps in concentrated therapeutic classes imply that the acquiring firm has profits that are at risk from the Arrow replacement effect, which has negative incentives for project continuation. Furthermore, preemption incentives imply that a firm with a product in a concentrated therapeutic class has an incentive to outbid rivals for acquisition targets that threaten the firm's existing business. Thus, incumbents are likely acquirers of promising R&D projects, but this research shows that they are less likely than other acquirers to move these acquired R&D projects to a commercial application. An implication is that dominant firms have incentives to use acquisitions to eliminate potential competition. Furthermore, Cunningham and the coauthors also conclude that the acquisitions observed in their study had a larger negative effect on R&D continuations than the benefits from diversification and synergies.[47]

These are important observations. Do they imply that antitrust policy should police acquisitions more aggressively? Yes, although it would be counter-productive to prohibit all acquisitions of projects that overlap with existing products or R&D projects. There is a trade-off between acquisitions that extinguish acquired R&D projects and acquisitions that promote incentives for innovation. R&D for buyout is a common business strategy in the pharmaceutical industry, as well as in other high-tech sectors. Some of the acquired projects will be terminated, but many others may not have existed but for the strategic option to sell them to an established firm. To avoid destroying incentives to create R&D projects in the first place, antitrust policy should focus enforcement on acquisitions of targets for which either a buyout is not a business strategy that motivates R&D effort or for which there are other acquirers that do not have significant competitive overlaps.

## 6   Divestitures and Compulsory Licensing

Antitrust authorities can address innovation concerns by pursuing divestitures or compulsory licensing of intellectual property, either as conditions to allow mergers or as remedies for monopolization. This section briefly reviews one of the most famous examples of antitrust divestiture—the breakup of American Telephone and Telegraph (AT&T)—and describes some notable cases of compulsory licensing to remedy monopolization. Chapter 7 reviews several merger cases in which enforcement agencies addressed innovation concerns by requiring divestitures of R&D assets or compulsory licensing obligations. The chapter reports that divestitures apparently resolved innovation concerns for some mergers, but divestitures were less successful in some other cases. Most of the compulsory licensing obligations had beneficial effects.

### The breakup of AT&T
In 1974, the Department of Justice (DOJ) filed an antitrust complaint against AT&T, alleging that AT&T had monopolized markets for a wide

range of telecommunications services and products in violation of Section 2 of the Sherman Act. AT&T (a.k.a. the Bell System) was fully integrated into all aspects of telecommunications services. It provided local exchange service through its Bell operating companies and long-distance service between the local exchanges. AT&T manufactured telecommunications equipment in its Western Electric subsidiary and owned Bell Laboratories, a research center with an illustrious history that produced a number of major discoveries, including the transistor, the laser, radio astronomy, information theory, and the Unix computer operating system.

The parties settled the antitrust charges in 1982 with a consent decree with the somewhat oxymoronic title of the Modification of Final Judgment (MFJ), which separated AT&T long distance services, Bell Labs, and Western Electric from the regional Bell operating companies. The MFJ also vacated a 1956 consent decree that restricted AT&T to the provision of regulated common carrier services and businesses incidental to furnishing those services.

The AT&T divestiture was a bold example of antitrust enforcement led by William Baxter, the Assistant Attorney General for Antitrust appointed by President Ronald Reagan. President Reagan worried that divestiture was too aggressive and told Baxter that, "When I was young, it cost 2 cents to mail a letter cross country and $2.00 to make a phone call. By the 1980s, each was 20 cents." Baxter reportedly replied, "Well, Mr. President, when I finish AT&T, I will be happy to take on the Post Office."[48]

The consent decree went into effect in 1984. The breakup had mixed, but overall positive, effects for innovation. Regrettably, the basic scientific research that had flourished at Bell Labs was a casualty of the breakup. The divestiture split Bell Labs into AT&T Bell Labs and Bell Communications Research (known colloquially as "Bellcore"), a smaller operation jointly owned by the operating companies. Bellcore was rebranded as Telcordia Technologies and was acquired by Ericsson in 2011. Most of Bell Labs went to Lucent in 1996, along with the remainder when Lucent and Alcatel merged in 2007. Nokia acquired Alcatel-Lucent in 2016, and what was left of Bell Labs was branded "Nokia Bell Labs." Ericsson and Nokia invest heavily in telecommunications R&D, but their focus is more applied than that of the historical research mission of Bell Laboratories.

Although the AT&T divestiture was a blow to basic scientific research and incurred large administrative costs, competition and

applied innovation thrived in the telecommunications industry following the breakup and further liberalization of the US telecommunications
industry by the 1996 Telecommunications Act. Prices for telecommunication services fell, as did prices for infrastructure and customer
premise equipment. Total R&D expenditures, while more focused on
applications, increased dramatically. New entrants provided a diversity of equipment choices and service options that did not exist under
the monolithic Bell System. The divestiture and subsequent liberalization of the Bell System demonstrate the positive benefits of competitive
markets for investment in applied R&D.

### Effects of compulsory licensing on industry innovation and price competition

As discussed in more detail in chapter 7, antitrust agencies have negotiated numerous consent decrees to settle competition concerns for proposed mergers and acquisitions, including concerns about harm to
innovation. These consent decrees often included requirements to license
intellectual property, along with other conditions. In the more distant
past, regulatory authorities, including antitrust agencies, have also used
compulsory licensing to address persistent monopolization. Because
licensing expands the universe of potential innovators, experience with
these obligations provides a window through which to examine the
effects of greater competition on innovation incentives.

In 1956, and long before the breakup of AT&T, the DOJ and AT&T
settled an antitrust complaint with a consent decree that obligated
AT&T to license its existing patents royalty-free and to license any
future patents at a reasonable royalty.[49] In the same year, the DOJ
settled antitrust charges against International Business Machines (IBM)
with a consent decree that included, among other provisions, the
requirement that IBM grant nonexclusive licenses royalty-free for any
of its patents existing at the time of the decree, plus offer licenses to
any patents filed during the subsequent five years at a reasonable
royalty.[50] Both decrees required applicants to grant reciprocal cross-
licenses to their patents. In 1975, the FTC resolved a complaint that the
Xerox Corporation monopolized the market for plain-paper office
copiers by requiring Xerox to grant licenses to its relevant patents to
any willing licensees at modest royalties.[51]

Compulsory licensing is controversial because it strips the patent
holder of rights that are bestowed by the patent system and can have
counterproductive consequences by weakening incentives to innovate.

The AT&T and IBM consent decrees were also controversial because they failed to undermine the monopoly power of these companies. AT&T maintained its grip on almost all aspects of telecommunications until it agreed to divestitures under the MFJ. IBM's control of the computer industry continued until technological advances in mini- and micro-computers chipped away at its mainframe monopoly. Only the Xerox consent decree truly opened the market to new competition, but some questioned whether the company engaged in conduct that warranted antitrust condemnation.[52]

Nonetheless, there is evidence that these consent decrees had beneficial effects for innovation and price competition. Martin Watzinger and coauthors rely on detailed information about citations to AT&T patents to conclude that the 1956 AT&T consent decree increased innovation that built on AT&T's patents. Specifically, they report that citations to AT&T's patents increased by 17 percent relative to similar patents in the same technology class that were not covered by the decree. The beneficiaries were mainly small and young companies that operated in fields other than telecommunications, which the authors attribute to the entry barriers that continued to protect AT&T from competition in regulated telecommunication markets. The authors also find that the positive effects for follow-on technologies from the decree more than compensated for a small reduction in patenting by AT&T.[53]

Others reached similar, if more qualitative, conclusions. Gordon Moore, a cofounder of Intel, credited the 1956 AT&T consent decree for much of the growth of the merchant semiconductor industry in the US.[54] AT&T was initially reluctant to license its fundamental patents on transistors, but it relented, in part to stave off antitrust actions.[55] David Mowery concludes that the 1956 AT&T consent decree supported high levels of knowledge diffusion and facilitated the entry of new competition in the semiconductor industry.[56] Peter Grindley and David Teece assert that "[AT&T's licensing policy shaped by antitrust policy] remains one of the most unheralded contributions to economic development—possibly far exceeding the Marshall plan in terms of wealth generation it established abroad and in the United States."[57]

Compared to the AT&T consent decree, available evidence on the benefits of compulsory licensing in the 1956 IBM consent decree is thin, perhaps because the patents at issue were not as fundamental as some of the patents owned by AT&T. However, there is no evidence that the 1956 IBM consent decree discouraged innovation. IBM introduced its enormously successful 360 series of mainframe computers in 1964 based on R&D efforts that occurred at the time of or soon after the decree. Furthermore, some commentators credit the decree and subsequent antitrust litigation with facilitating the entry of plug-compatible peripherals for IBM mainframes and for enabling the development of competing mainframe computers sold by Amdahl and Control Data.[58]

The 1975 Xerox consent decree is noteworthy because, to quote Willard Tom, "Wrong-headed as much of the case appears today, it seems to have done a world of good."[59] The consent decree was "wrong-headed" in the sense that it settled numerous allegations, such as discriminatory pricing arrangements, that draw little antitrust scrutiny today. But the decree accomplished "a world of good" by lowering prices and increasing the diversity of plain-paper copiers for businesses and consumers.[60]

Although IBM and Litton were successful entrants into the market for plain-paper copiers prior to 1975, Xerox's broad patent portfolio helped to protect Xerox from new competition. After the decree was entered, there was innovation and new competition in plain-paper copiers from manufacturers of low-volume, coated-paper copiers and manufacturers of high-volume photographic machines, including Canon, Toshiba, Sharp, Panasonic, Konica, and Minolta. The decree did a "world of good" for consumers, but not for Xerox. Xerox's market share plummeted following entry by new rivals. Xerox executives criticized the FTC compulsory licensing decree for exposing the company to foreign competition, but they also acknowledged that competition forced Xerox to successfully reinvent itself and become a more efficient company.[61]

Other studies confirm increases in the pace of innovation following events that make patented technologies available for use at zero or low royalties. Alberto Galasso and Mark Schankerman examine the effects of litigation outcomes that invalidate patents, thereby making the technologies covered by the patents available for use without compensation, which is similar in some respects to royalty-free compulsory licensing. Citations increased by an average of 50 percent following a finding of invalidity, with most of the effect concentrated in the fields of computers and communications, electronics, biotechnology, and medical instruments.[62] A finding of invalidity only increased citations when the patent was owned by a large firm, with the increase in citations coming mostly from small innovators.

Petra Moser and Alessandra Voena explore the consequences of an episode of compulsory licensing in the US that occurred during and after World War I under the Trading with the Enemy Act.[63] The Act initially permitted US firms to practice enemy-owned patents if they contributed to the war effort. Congress amended the legislation in 1918 to confiscate all enemy-owned patents, and by 1919, German-owned patents were systematically licensed without compensation to US firms. Moser and Voena find that the number of patents granted to US inventors in technological fields with at least one confiscated German patent increased by an average of about 20 percent relative to patents granted to US inventors in similar fields that did not have confiscated patents. Subsequent research shows that the confiscation of intellectual property did not inhibit innovation by the firms whose property was confiscated. The German firms increased their investment in R&D and generated 30 percent more patents after the 1918 confiscation.[64]

Studies of compulsory licensing obligations conducted by Michael Scherer validate the benefits of compulsory licensing for follow-on innovations and do not support a general conclusion that compulsory licensing harms innovation incentives for the firms that were subjected to these requirements.[65] Examples of merger consent decrees that compel the licensing of intellectual property reviewed by Colleen Chien and the examples reviewed in chapter 7 reach similar conclusions.[66]

These studies and observations from the AT&T, IBM, and Xerox consent decrees find that making patented inventions available for use royalty-free or at reasonable royalties has a positive impact on follow-on innovations that build on the patented inventions. While innovators clearly need some protection from imitation, there is little evidence that the largely unanticipated weakening of patent rights of the type encountered in a few compulsory licensing decrees has had offsetting negative effects on invention. The benefits of compulsory licensing stand in stark contrast to arguments that compulsory licensing obligations would strike the death knell for innovation by undermining incentives from intellectual property protection.[67] The evidence is that compulsory licensing, when selectively applied, can promote innovation, in addition to having salutary effects by addressing competition concerns for mergers and limiting industry dominance.

Even if compulsory licensing has a deterrent effect on some innovations, there is good reason to conclude that a slight weakening of patent rights has a positive effect on industry innovation and competition. As discussed in chapter 4, a reduction in royalties charged by frontier innovators or an increase in access to these technologies can increase economic welfare by promoting follow-on innovation, which can be accomplished with a compulsory licensing obligation at a reasonable royalty.

## 7   Concluding Remarks on Empirical Evidence

Empirical research on the link between competition and innovation has evolved over many decades. Early investigations found that measures of innovation peaked at intermediate levels of industry concentration. However, these studies failed to account for industry factors and technological characteristics that affect innovation opportunities and the ability to appropriate returns from R&D investments. More recent interindustry studies attempt to account for these factors and apply best statistical practices to control for spurious correlations. Several of these studies find a positive relationship between competition and innovation. Other studies find that moderate levels of industry competition are more conducive to innovation, and there is some contrasting support for a negative relationship.

Empirical economic research has yet to reach a consensus about the interaction between market competition and innovation incentives, although the evidence does clearly highlight the significance of technological asymmetries for firms' innovation incentives. Competition appears to have a much greater beneficial effect on R&D investment and innovation by firms that are close to the industry's technological frontier. This is consistent with the theoretical model of stepwise innovation. An increase in competition makes it more profitable for firms that are close to the industry technological frontier to escape competition by innovating. On the other hand, competition can deter innovation for firms that are far from the frontier by making it more difficult for these firms to profit from catching up to their more advanced rivals.

Antitrust enforcers often presume that mergers that substantially lessen competition in a high-tech industry are likely to harm innovation. Unfortunately, there is little empirical research that proves this presumption. Empirical analysis of the effects of mergers on innovation is challenging because it is difficult to find a natural experiment in which a merger affects innovation incentives differently for some set of firms compared to other, otherwise similar, firms. Furthermore, data is rarely available to study the effects of mergers in highly concentrated industries because antitrust authorities typically prevent mergers in highly concentrated industries or condition them on remedies that are intended to eliminate alleged anticompetitive effects. Nonetheless, available evidence does not support a conclusion that mergers generally promote innovation or competition for future products or services.

More aggressive antitrust enforcement can prevent so-called killer acquisitions, in which an established firm acquires and then extinguishes a promising R&D project. Empirical evidence on thousands of acquisitions in the pharmaceutical industry demonstrates that an acquiring firm is more likely to terminate a research project if it is in a concentrated sector and the acquired project is a potential competitor to its products or research projects that are directed to similar therapeutic applications. However, a prohibition on such acquisitions could have negative repercussions by destroying incentives for investment in R&D by entities that rely on acquisitions to monetize their R&D efforts. A better policy would encourage alternative acquisitions by firms that do not have significant product or technology overlaps.

Evidence from compulsory licensing obligations supports the conclusion that opening markets to competition by making intellectual property more generally available has beneficial effects for follow-on innovation and price competition. Widespread compulsory licensing would have negative consequences for incentives to invest in fundamental innovations, but the evidence does not show adverse effects from the few episodes in which antitrust authorities and regulators have used compulsory licensing to open markets to competition. The next chapter explores several examples of divestitures and compulsory licensing that have addressed innovation concerns for mergers.

# 7  Merger Enforcement for Innovation: Examples and Lessons for Remedies

Getting merger remedies right is central to merger enforcement policy. ... A remedy should fully and squarely cure the violation. It needs to preserve the status quo ante in affected markets by effectively addressing any and all anticompetitive effects arising from the transaction.
—Acting Associate Attorney General Bill Baer, Remarks at the American Antitrust Institute's 17th Annual Conference, Washington, DC (2016)

## 1  Introduction

Courts in the US and Europe have yet to litigate antitrust cases that solely allege innovation effects, but antitrust authorities in these and other jurisdictions have challenged numerous mergers that threaten innovation. This chapter reviews selected merger cases for which antitrust enforcement likely protected innovation incentives and future competition, as well as some cases that were apparently less successful in this respect.

Parties have abandoned proposed mergers in response to agency concerns about innovation effects. In most cases, the agencies have allowed mergers to proceed with consent decrees. This chapter classifies merger challenges alleging innovation harms into three broad categories:

- *Unconditional challenges:* These are challenges in which the merging parties and the reviewing agency cannot agree to a remedy to address alleged harms and the agency acts to block the merger. Parties often abandon a proposed merger following an unconditional challenge. Some parties have pursued litigation to adjudicate an unconditional challenge, although no court has yet litigated a merger challenge based solely or primarily on alleged harms to innovation.

• *Structural remedies:* These are consent decrees that require the merging parties to divest assets to remedy lost competition. The required divestitures may include intellectual property, know-how, and other resources in addition to physical assets.

• *Behavioral remedies:* These are consent decrees that impose restrictions on the merged entity, but do not require asset divestitures. Although behavioral remedies may encompass a wide range of activities, the focus in this chapter is on compulsory licensing obligations that require the merging parties to grant nonexclusive licenses to patents or other intellectual property with no or low royalties.[1]

Several studies have examined whether merger approvals that were conditioned on structural or behavioral remedies succeeded in preventing postmerger price increases for existing products or services.[2] This chapter deals with merger conditions that involve research and development (R&D) assets and related intellectual property to address concerns about innovation and future price competition. The lessons that can be drawn from these examples are limited because the innovations that would have occurred absent antitrust intervention cannot be established with certainty. Nonetheless, they reveal some useful patterns.

Section 2 of this chapter begins by reviewing a selection of unconditional merger challenges by US antitrust authorities. In response to these unconditional challenges, the parties abandoned their proposed transactions and continued to innovate successfully as separate entities. The implication is that antitrust authorities should not be easily swayed by testimony that a merger is essential to create incentives for innovation. Section 3 also discusses a transaction that the reviewing agency did not challenge, and which had significant implications for innovation.

Section 4 reviews examples of structural remedies to address innovation concerns from mergers. Targeted R&D divestitures facilitated innovation in many cases, but the divestitures failed to restore innovation incentives in other cases. Some acquirers of divested assets did not invest in the relevant R&D or encountered financial difficulties. Several firms that acquired divested assets subsequently merged or were acquired, and their successors did not continue to pursue relevant R&D. In some cases, innovations came from unexpected sources, which negated the importance of the divestiture.

Asset divestitures are tricky instruments for resolving innovation concerns. The key R&D assets are scientists and engineers who cannot be compelled to move to a different company. Acquiring companies may choose to purchase divested R&D assets as a relatively inexpensive option to explore the potential of a technology, or to acquire intellectual property and human capital that can be useful in other applications.

Examples of disappointing innovation results following some asset divestitures are not sufficient to conclude that structural remedies are ineffective merger enforcement tools for innovation. One cannot be certain that unrealized expectations were the fault of poorly designed divestitures or are more properly attributed to the vagaries of innovation. The results from these examples should be interpreted with appropriate caution, but they establish a need for in-depth studies of the consequences from merger remedies intended to address alleged harm to innovation.

Section 5 reviews several examples of merger remedies that involved compulsory licensing obligations. In contrast to the evidence regarding structural remedies, the examples of compulsory licensing appear to demonstrate more consistent benefits. Of course, these results reflect the circumstances in which agencies pursued structural or behavioral remedies. The merger cases that were conditioned on structural remedies may have presented innovation concerns that were particularly difficult to address, and it does not follow that compulsory licensing would have been a more effective remedy in these situations.

Nonetheless, the evidence is that compulsory licensing of intellectual property rights can be an effective merger remedy in some circumstances. Although compulsory licensing can undermine incentives for breakthrough innovations if it allows licensees to copy inventions at low cost, that is typically not the case. Case studies and academic research indicate that most instances of compulsory licensing have benefited follow-on innovation and price competition, with few offsetting disincentives for R&D.[3]

Section 6 turns to several examples of merger challenges by the European Commission (EC). The EC was a late entrant into merger enforcement for innovation, with its first challenge occurring several years after complaints in the US that alleged innovation harm from mergers. Nonetheless, it made up for its slow start with several challenges that addressed innovation concerns, including at least one challenge that reflects a more aggressive enforcement posture for alleged innovation harm from a merger than in any case yet pursued by US antitrus enforcers. The final section concludes with lessons that the examples reviewed in this chapter suggest for merger enforcement policy for innovation.

## 2   Unconditional Merger Challenges

Sometimes antitrust authorities conclude that a merger threatens innovation and that their concerns cannot be remedied with piecemeal
commitments. In these circumstances, the agencies must choose to
either allow a merger that raises innovation concerns with no conditions or challenge the merger without conditioning the transaction on
structural or behavioral commitments.

**Innovation shifts into high gear:**
**General Motors—ZF Friedrichshafen**
Chapter 1 of this book began with a 1993 case in which the US Department of Justice (DOJ) blocked the proposed acquisition of the Allison
Transmission Division of General Motors (GM) by ZF Friedrichshafen
AG ("GM-ZF"). The complaint identified an "innovation market" for
"technological innovation in the design, development, and production
of medium and heavy automatic transmissions for commercial and
military vehicles" and further explained that "the proposed transaction
will reduce the number of competitors in the innovation market from
three to two, reducing both the actual competition for innovation and
the incentive of the remaining firms to compete vigorously for future
innovation."[4]

The GM-ZF complaint provides a blueprint to construct an allegation of harm to innovation from overlapping R&D capabilities, specifically by: defining the relevant innovation market (which the US antitrust
agencies now term a "research and development market"); showing
that few firms possess the specialized assets necessary to compete in
the market; and citing evidence that the acquisition or merger would
be likely to reduce the firms' incentives to invest in new product development, or cause a loss of future price competition. The decision to
block the merger was apparently sound. Competition and innovation
subsequently thrived in markets for medium and heavy-duty automatic transmissions for commercial and military vehicles.

**Two hearts are better than one: Thoratec—HeartWare**
Heart failure is one of the leading causes of death in the developed
world. Patients who suffer from end-stage heart failure are possible
candidates for surgically implanted mechanical blood pumps, for
which left ventricular assist devices (LVADs) are the most common.
The Thoratec Corporation was a leader in this technology at the begin-

ning of the millennium. Thoratec sold the HeartMate XVE and a second-generation product, the HeartMate II. At the time, these were the only LVADs approved by the US Food and Drug Administration (FDA) for destination therapy, rather than as a bridge to a transplant. HeartWare was a potential new entrant into the line of LVADs with a device that it called the HeartWare HVAD.

In February 2009, Thoratec proposed to purchase HeartWare, in a transaction valued at $282 million.[5] HeartWare's HVAD was in clinical trials at the time of the proposed acquisition, and initial results were promising. Doctors praised the HVAD's innovative design and small size, which simplified surgical implantation. Nonetheless, whether HeartWare's HVAD would prove to be superior to Thoratec's LVADs was uncertain. The high mortality rate for patients with end-stage heart failure and the difficulty of using mechanical devices to replace or assist normal heart functions complicate evaluations of clinical trials and comparisons with other devices.

The Federal Trade Commission (FTC) investigated and concluded that the merger, if permitted, would eliminate current and future competition between Thoratec and HeartWare, deny life-sustaining treatments for end-stage heart failure patients, force them to pay higher prices, and eliminate innovation competition. The FTC formally challenged the merger in July 2009, and the parties abandoned the transaction soon afterward.[6]

Did the FTC make the right call? The merger raised clear concerns about harm to product-to-project innovation competition of the type described in chapter 5. Prior to the merger, Thoratec was the only LVAD supplier for a therapy for which there are few if any alternatives. LVADs are used only after all other potential treatments, including drugs, surgery, and other medical devices, have been exhausted. HeartWare promised to be a significant new competitor in this market. Price competition is less important for LVADs than for many other products, but innovation competition is critical and could have been compromised if Thoratec controlled all FDA-approved LVADs.

Had Thoratec challenged the FTC decision, the company might have made several counterarguments:

• HeartWare's LVAD was unproven, and there was a high probability at the time of the merger that the HVAD would not be an acceptable device; that is, there was a high probability of "no harm, no foul."
• Other LVADs were in the early stages of clinical trials.

- The merged company could exploit synergies in the development of new LVADs.
- Blocking Thoratec's purchase of HeartWare would be detrimental for innovation because firms such as HeartWare invest millions of dollars in new technology with the expectation of selling to established firms like Thoratec.

The first counterargument is not compelling. As discussed in chapter 5, if there are no merger-specific efficiencies, the decision to challenge the acquisition of a potential competitor should not hinge on the probability that entry will be successful. If Thoratec's acquisition of Heart-Ware would extinguish or significantly harm innovation and price competition that would have occurred if HeartWare had remained independent, then antitrust enforcement should prevent such an outcome even if the innovation has a low probability of success, provided that the acquisition does not create offsetting efficiency benefits.

The second counterargument raises an empirical question. If a large number of LVADs were likely to enter the market as new, safe, effective, and economic alternatives, then the elimination or suppression of one LVAD would not likely have significant consequences for competition. In fact, there were only a few other LVADs in development when Thoratec offered to acquire HeartWare, and as of 2019 the FDA had approved only one other LVAD for adult destination therapy, in addition to the LVADs sold by Thoratec and HeartWare.

The third counterargument is a merger-specific efficiencies defense. Thoratec might have argued that it would devote more financial resources to develop the HVAD than HeartWare could profitably attract and invest as an independent concern, or that the combination of the HVAD and Thoratec's other LVADs would create complementary therapeutic possibilities that would not be available without the merger. Because these are merger-specific defenses, the burden is on the merging parties to demonstrate their validity.

The fourth counterargument raises a more general issue, also discussed in chapter 5. Antitrust enforcement has to assess the risk of allowing acquisitions that extinguish innovation and future competition, while also allowing opportunity for innovative firms to pursue buyouts as exit strategies. The FTC appeared to get the balance right in Thoratec-HeartWare. Thoratec and HeartWare remain two of a very few entities that have won FDA approval to sell (or even research) LVADs. In 2016, the medical device giant Medtronic purchased Heart-

Ware in an acquisition valued at $1.1 billion, almost four times the amount that Thoratec agreed to pay for HeartWare in 2009. St. Jude Medical acquired Thoratec in 2016 and was later acquired by Abbott Laboratories. As subsidiaries of these larger entities, Thoratec and HeartWare continue to innovate in this area. Thoratec recently developed the HeartMate III, which is smaller than the HeartMate II and has a magnetically levitated rotor. HeartWare supported clinical trials for a new LVAD that is one-third the size of its already small HVAD device.

**Innovation at the leading edge: Applied Materials—Tokyo Electron**
The proposed merger of Applied Materials and Tokyo Electron in 2015 is another example of a transaction for which the Antitrust Division of the DOJ concluded that there were no acceptable remedies for the predicted harms to innovation and future price competition. The firms abandoned their proposed merger after the Division informed them of its competition concerns.

Applied Materials and Tokyo Electron are two of very few firms with the capability to develop and manufacture leading-edge semiconductor tools for high-volume semiconductor manufacturing. The Division identified narrow overlaps in existing product markets and pipeline projects that raised concerns about product-to-project competition. It also emphasized broader concerns related to the parties' differential capabilities to develop future high-value manufacturing tools for the semiconductor industry.[7]

Nicholas Hill, Nancy Rose, and Tor Winston, economists in the Antitrust Division involved in the review of the proposed merger, wrote, "Because AMAT and TEL are so capable, they are often the two best (or among the three best) development partners to solve a leading-edge semiconductor manufacturer's high-value deposition and etch problems. The merger would have eliminated the competition between AMAT and TEL to be selected as a future development partner, as well as any eventual competition between their competing products."[8] They added that "replicating the competitive significance of one of the most innovative companies in a sector that is virtually synonymous with innovation would be exceptionally challenging."[9] The Division rejected proposed remedies because the necessary assets to address future innovation concerns could not be isolated from the companies' broader capabilities and experiences with material sciences and engineering.

Applied Materials and Tokyo Electron continued to invest in R&D following their attempted merger. Table 7.1 shows the companies' R&D

Table 7.1
Annual research, development, and engineering expenses (merger proposed in 2015).

|                                  | 2013  | 2014  | 2015   | 2016  | 2017  | 2018  |
|----------------------------------|-------|-------|--------|-------|-------|-------|
| Applied Materials ($ millions)   | 1,320 | 1,428 | 1,451  | 1,540 | 1,774 | 2,019 |
| Tokyo Electron ($ millions)*     | 664   | 718   | 645    | 694   | 764   | 883   |
| AMAT + TEL                       | 1,984 | 2,146 | 2,096  | 2,234 | 2,538 | 2,902 |
| Percent change in AMAT + TEL     |       | 8.2%  | (2.3%) | 6.6%  | 13.6% | 14.3% |

* Converted from yen at 110 yen/dollar
Source: Applied Materials and Tokyo Electron Annual Reports.

expenditures for the years 2013–2018. Both companies increased their R&D substantially in the two years following their attempted merger compared to the previous two-year period. Of course, table 7.1 does not establish that the companies would have invested less in R&D if they had merged. Moreover, as I have stressed elsewhere in this book, R&D is an input to innovation and not a measure of innovation output. It is possible that the combined companies could have achieved the same or greater innovation with less R&D expense. Nonetheless, the evidence suggests a continued commitment to innovation by both companies in the two years following their attempted merger. Applied Materials reported that the company "continued to prioritize existing RD&E investments in technical capabilities and critical research and development programs in current and new markets, with a focus on semiconductor technologies."[10] Tokyo Electron emphasized its aim to increase profitability and market share with its commitment to technological differentiation.[11]

A year later, two other semiconductor equipment companies, Lam Research Corp. and KLA-Tencor Corp., abandoned their plans to merge after the DOJ informed the companies that it had serious concerns about the effect of the proposed transaction on innovation. Lam Research is a leading provider of etch, deposition and clean tools and process technology used in the fabrication of semiconductors. KLA-Tencor is the leading provider of semiconductor fabrication metrology and inspection equipment.

Unlike the proposed merger of Applied Materials and Tokyo Electron, this was a merger of firms that operate in complementary markets. Firms that supply complementary fabrication tools can coordinate prices and product development to benefit semiconductor manufacturers.[12] Nonetheless, the Antitrust Division opposed the merger because it "presented concerns about the ability of the merged firm to foreclose competitors' development of leading edge fabrication tools and process technology on a timely basis."[13] The Division did not explain why the merged firm would have a greater incentive to foreclose competitors than either party would have acting individually. Apparently, a concern was that by acquiring KLA-Tencor, Lam Research would obtain information about competitors' technologies that it could exploit for strategic advantage.[14]

The Division's opposition to this merger contrasts with other merger enforcement actions in which the Antitrust Division or the FTC accepted behavioral commitments for suppliers of complementary products.[15] The FTC approved Broadcom's acquisition of Brocade Systems subject to a requirement that Broadcom implement firewalls to protect confidential business information.[16] Brocade manufactures channel switches used to direct traffic on a fiber communications network. Broadcom makes application-specific integrated circuits (ASICs) for fiber switches. Broadcom supplies ASICs to Cisco, which is Brocade's main competitor in the market for fiber channel switches. A key concern was that Broadcom's relationship with Cisco would provide competitively sensitive confidential information, which the merged company could use to harm competition for fiber channel switches. The parties accepted a consent decree that required Broadcom to implement firewalls preventing the flow of Cisco's confidential business information outside of an identified group of relevant Broadcom employees.

Antitrust enforcement is a fact-specific process and it is not unusual for enforcement decisions to differ for cases that appear to be similar; however, the antitrust authorities would do a service to explain why they reach different outcomes in otherwise similar cases. It is unclear why the FTC believed that firewalls would remedy the potential for competitive harm if Broadcom misused Cisco's competitively sensitive confidential information while the Antitrust Division could find no acceptable remedy for the merger of Lam Research and KLA-Tencor.

In 1997, the DOJ blocked the proposed merger of Lockheed Martin and Northrop Grumman. The complaint cited the loss of product-to-project competition and project-to-project competition of the types described in chapter 5, and the loss of price competition for several high-tech systems, including high-performance fixed-wing military aircraft, airborne early warning radar, infrared countermeasures, anti-submarine warfare systems, and missile warning systems.[17] A year later, the FTC approved the merger of Boeing and McDonnell Douglas, the two largest US manufacturers of commercial aircraft. Although the Commission did not contrast its decision with the DOJ decision in Lockheed Martin–Northrop, it explained that McDonnell Douglas had fallen too far behind the frontier of commercial aircraft technology to pose meaningful competition.[18]

There are many other examples of mergers that the antitrust agencies blocked after citing significant innovation concerns. In 2002, the FTC blocked the proposed merger of Cytyc and Digene, both of which developed and sold products used to screen for cervical cancer.[19] In 2009, the DOJ litigated the completed merger of Microsemi Corp. and Semicoa, Inc., manufacturers of small signal transistors and ultrafast recovery rectifier diodes used in critical military and civil applications. This caused Microsemi to reverse the merger and divest the Semicoa assets in their entirety.[20]

Both cases alleged a merger to monopoly with adverse effects for prices and innovation. The parties involved in both mergers continue to innovate. Cytyc claims that its ThinPrep system is the most widely used method for cervical cancer screening in the US,[21] and Digene asserts that it is a leader in molecular diagnostics, and that it develops, manufactures, and markets proprietary deoxyribonucleic acid (DNA) and ribonucleic acid (RNA) tests focused on women's cancers and infectious diseases.[22] Both Microsemi and Semicoa advertise high-reliability, military-grade semiconductor devices.[23]

## 3   A Notable Transaction Not Challenged

The preceding section identified several proposed mergers and acquisitions that the antitrust agencies challenged, in part because they threatened harm to innovation and there were no mutually acceptable mitigating conditions. The antitrust agencies also have taken innovation into account in decisions not to challenge mergers in highly concentrated industries. An example is the acquisition of Novazyme Pharmaceuticals by Genzyme. Genzyme and Novazyme were the only two companies with active research programs for Pompe disease, a rare genetic, and often fatal, muscular disorder.

Genzyme acquired Novazyme in September 2001. The FTC subsequently reviewed the transaction and chose not to take any enforcement action. In a lengthy press release, Chairman Timothy Muris argued that there was no presumption that a merger, even a merger to monopoly, would harm innovation competition; he also pointed to the exceptional factual circumstances of the case.[24] John Crowley was chairman and a cofounder of Novazyme. Two of his children had been diagnosed with Pompe disease, and Crowley was determined to find a cure.[25]

Genzyme terminated the Novazyme project not long after the acquisition. It might appear that the FTC made the wrong call in its decision not to challenge this acquisition because it created a monopoly in a R&D market for Pompe disease therapies, and only one R&D project survived the acquisition. However, the facts of the case suggest that Genzyme's decision to terminate the Novazyme project was the result of the challenges to find a cure for Pompe disease, and it is unlikely that Novazyme would have survived as an alternative provider of a therapy for Pompe disease if the company had remained independent.

Novazyme was far behind Genzyme at the time of the acquisition. If Novazyme had remained an independent concern and produced a drug after Genzyme, Novazyme would have faced a high barrier to challenge Genzyme's exclusivity. Pompe disease qualifies for exceptional treatment under the Orphan Drug Act, which guarantees seven years of exclusivity for a new drug to treat diseases that affect small populations unless another drug can be proved to be superior in head-to-head clinical trials. Furthermore, the merger had credible, merger-specific synergies for R&D. Genzyme had experience with enzyme replacement therapies and had the manufacturing capacity to produce a drug, while Novazyme had neither. The Genzyme-Novazyme case validates the principle that facts are critical to competition analysis.

## 4   Structural Merger Remedies for Innovation

The DOJ and FTC raised innovation concerns in more than 100 cases since 1995. In most of these cases, the reviewing agency allowed the merger to proceed after the parties agreed to specific conditions, such as the licensing of intellectual property or divestiture of assets, including assets necessary for effective R&D.

The objective of a remedy is to replace the competition lost by a merger or acquisition. This is a challenging task for innovation competition. The recipient of divested R&D assets or intellectual property may be unable to replace lost innovation because: it does not possess necessary resources for effective R&D; it has strategic objectives that differ from investing in the technological space for which competition was lost from the merger; or the relevant scientists and engineers choose not to move to the firm that acquired the assets or intellectual property. In their study of more than 8,000 acquisitions of new drug R&D projects, Colleen Cunningham and the coauthors find that only 22 percent of preacquisition inventors moved to the acquirer after the acquisition.[26]

Despite these obstacles, many examples of R&D divestitures appear to demonstrate successful restoration of innovation incentives allegedly threatened by mergers. The divestiture of Covidien's R&D assets for drug-coated balloon angioplasty to treat cardiovascular disease to the Spectranetics Corporation restored innovation competition threatened by Medtronics's acquisition of Covidien. The divestiture of Novartis oncology compounds to Array BioPharma remedied innovation concerns caused by the acquisition of GlaxoSmithKline's (GSK's) oncology portfolio by Novartis.

There are many other cases of divestitures of R&D assets that appear to have successfully restored innovation incentives that were threatened by a merger. The successful cases are generally those for which: the divested assets are well-defined projects in a late stage of development; divested assets include related intellectual property, specialized manufacturing equipment, and key personnel; and the recipient of the divested assets is a firm with a strong track record in the relevant R&D activities.

When these conditions are absent, R&D divestitures required in merger consent decrees have sometimes failed to deliver hoped-for innovation. I review several of these cases in this section.[27] They suggest that piecemeal divestitures may not resolve innovation concerns. An alternative would be to challenge the merger unconditionally. However, by drawing a hard line for mergers that may threaten innovation, the antitrust agencies may incentivize merging parties to divest troublesome assets before they file for merger approval or during the merger review. There is no reason to believe that this "fix it first" strategy would be any more effective at restoring innovation competition than would a remedy that is negotiated with the reviewing antitrust authority. Alternatively, the agencies should focus on compulsory licensing obligations, which as section 5 shows, have often had procompetitive benefits.

I caution that ex post studies of market performance provide only a narrow view of the effectiveness of consent decrees to remedy innovation harm from a merger because innovation is inherently uncertain. Lack of future competition from a firm that acquires divested assets or intellectual property licenses can reflect lack of commercial success and need not be indicative of a failure to invest in R&D.

### American Home Products–American Cyanamid (1995)

American Home Products (AHP) and American Cyanamid agreed to merge in August 1994. Among other activities, both companies manufactured and sold vaccines for tetanus and diphtheria and had active research programs for a vaccine to treat Rotavirus infections, a common cause of severe diarrhea. Dehydration is a serious complication of Rotavirus and is a major cause of childhood deaths in developing countries.

As a condition to approve the merger, the FTC approved the application of AHP to divest Cyanamid's intellectual property concerning the Rotavirus vaccine research and related assets and information to Green Cross Korea, which was active in vaccine research and manufacturing.[28] Green Cross Korea later rebranded itself as GC Pharma. In 2002 AHP changed its name to Wyeth, which Pfizer purchased in 2009.

To date, neither Green Cross Korea (or GC Pharma), AHP, Wyeth, nor Pfizer has developed an FDA-approved vaccine to protect humans against Rotavirus disease. In 2006, the FDA approved RotaTeq, manufactured by Merck, for the prevention of Rotavirus gastroenteritis in infants six weeks to thirty-two weeks of age. In 2009, the FDA approved Rotarix, manufactured by GSK, for the prevention of rotavirus gastroenteritis for use in infants six weeks to twenty-four weeks of age.

There is no evidence that Green Cross Korea (or GC Pharma) made a determined effort to develop a Rotavirus vaccine. The company did not apply for any US patents that reference "Rotavirus" in the two decades following the merger. GC Pharma currently offers a vaccine for chickenpox and has vaccines in various stages of investigation and clinical trials for several diseases and maladies, but nothing for Rotavirus.[29]

### Glaxo Wellcome–SmithKlineBeecham (2000)

The pharmaceutical companies Glaxo Wellcome (GW) and SmithKlineBeecham (SB) announced a proposed merger in January 2000, which would form GSK in a transaction then valued at about $182 billion. Prior to the merger, GW had a cooperative relationship with Gilead Sciences to develop a topoisomerase I inhibitor for the treatment of certain cancers. SB already had sold Hycamptin, the leading drug in this category. The FTC reviewed the proposed merger and conditioned approval on a requirement that the merged company divest its rights to the topoisomerase I inhibitor under development at GW to Gilead Sciences.[30]

GW and SB were also two of only a few firms developing prophylactic vaccines to prevent herpes infection. GW was relatively new to vaccine research, but had a significant development project underway to develop vaccines against genital herpes using a technology developed by Cantab Pharmaceuticals. To remedy concerns about R&D competition in this space, the Commission required the merged company to return to Cantab all rights, information, and results from clinical trials that would be necessary for Cantab to develop a prophylactic herpes vaccine.[31]

Neither of these actions resulted in new drug competition. Gilead sold its oncology assets (including the topoisomerase I inhibitor project) to OSI Pharmaceuticals in November 2001,[32] less than a year after the Commission approved the merger. OSI continued R&D on the topoisomerase I inhibitor project for a time, but it terminated the project in 2004 because it concluded that clinical results did not sufficiently differentiate the program from Hycamptin sold by GSK.[33]

Several firms have clinical trials under way to evaluate a prophylactic herpes vaccine, but there are no prophylactic vaccines on the market to date. The drug that Cantab was developing in conjunction with GW failed phase II clinical trials in 2002.[34] Cantab went into a financial tailspin and was acquired by Xenova Pharmaceuticals in the same year.[35] Celtic Pharmaceuticals acquired Xenova in 2005.[36] GSK had a therapeutic herpes vaccine project under development, but it failed to show efficacy in phase III clinical trials, and the company terminated the program in 2010.[37] A search of clinicaltrials.gov did not identify any ongoing trials for a herpes vaccine sponsored by Cantab, Xenova, or Celtic Pharmaceuticals.

### Heraeus Electro-Nite–Midwest Instrument Co. (2014)

Heraeus Electro-Nite acquired Midwest Instrument Company (Minco) in 2012. Both parties sold single-use sensors and instruments used to measure and monitor the temperature and chemical composition of molten steel. The transaction initially escaped antitrust review because it was below the Hart-Scott-Rodino (HSR) thresholds that require notice of mergers and acquisitions to the FTC.

The DOJ caught wind of the acquisition from complaining customers. In 2014, the DOJ filed a complaint alleging elimination of competition in the development, production, sale, and service of sensors and instruments for steelmaking in the US, with consequences for higher prices and reduced innovation.[38] The DOJ required Heraeus to divest assets and intellectual property acquired from Minco.[39] The designated acquirer was a new entrant, Keystone Sensors LLC, formed in 2013 for the purpose of entering the US market for sensors and instruments to provide an alternative to Heraeus. Unfortunately, Keystone announced plans to reorganize on April 30, 2016, and has not been heard from since.

The DOJ almost certainly would have challenged the transaction if it had been notified of it because it was essentially a merger to monopoly. At the time of the acquisition, Heraeus accounted for about 60 percent of sensor and instrument sales, with almost all of the remainder sold by Minco. Forced to act after the acquisition was concluded, the only alternatives available to the DOJ were either to do nothing or structure a divestiture. The DOJ chose the latter course, but its experience demonstrates the difficulty of replacing lost competition and innovation.

### Nielsen Holdings–Arbitron (2014)

The FTC reviewed a proposal by Nielsen Holdings to acquire Arbitron in 2014. Both companies were in the business of providing audience measurement services. Audience measurement services provide useful metrics for advertisers and media networks in negotiations over purchases and sales of commercial airtime.

Nielsen is the dominant provider of television audience measurement services in the US. In 2014, Arbitron had a comparable position as a provider of radio measurement services. The FTC alleged that both companies were well positioned to provide national syndicated cross-platform measurement services. A syndicated cross-platform audience measurement service accounts for audience participation across multiple media platforms, including online and mobile platforms in addition to television and radio, and offers the data to subscribers. Demand for such a service was increasing rapidly at the time of the proposed acquisition, along with the profusion of various media platforms.[40]

Television is a critical component of cross-platform measurement services, and the FTC concluded that only Nielsen and Arbitron maintained large, representative panels capable of measuring television with the required individual-level demographics. Other firms were not as well positioned to compete with Nielsen and Arbitron to develop a national syndicated cross-platform audience measurement service because they lacked the representative panels, existing audience measurement technology assets of the quality and character of Nielsen's and Arbitron's, and strong brands in audience measurement.[41]

The proposed acquisition of Arbitron by Nielson posed a threat to project-to-project innovation competition of the type described in chapter 5. Each company had a national syndicated cross-platform project in its R&D pipeline. The FTC alleged that without conditions, the acquisition would eliminate future competition and result in less innovation for national syndicated cross-platform audience measurement services.

The Commission conditioned approval of the acquisition on the divestiture of assets related to Arbitron's cross-platform audience measurement business, including data from its representative panel. The consent agreement also required Nielsen to provide the acquirer of these assets with a perpetual, royalty-free license to data, including individual-level demographic data, and technology related to Arbitron's cross-platform audience measurement business for a period of no less than eight years. In addition, the Commission required Nielsen to make improvements and enhancements to the Arbitron panels at the request and expense of the acquirer in order to further its ability to offer a national syndicated cross-platform audience measurement service.

Comscore Inc. acquired the relevant assets. Comscore appeared to be a suitable replacement for competition lost from the acquisition. Arbitron had previously partnered with Comscore to provide customized cross-platform audience measurement services to ESPN and both companies were developing national syndicated cross-platform offerings.[42]

As of 2018, both Nielsen and Comscore purport to offer national syndicated cross-platform audience measurement services, but the services are in a nascent state.[43] Comscore has had serious organizational problems,[44] and the two companies have been mired in litigation over their interpretations of commitments from the consent decree.[45]

It is not obvious that an acquisition without the required divestitures would have had a worse outcome. Television and radio audience measurement services are complementary components of a cross-platform audience measurement service. The FTC apparently discounted the possibility that a merged Nielsen and Arbitron would be a more effective innovator by exploiting their complementary assets without a divestiture to Comscore. The resulting experience does not inspire a lot of confidence that the FTC's prediction has been realized.

Targeted divestitures can successfully restore competition and innovation incentives when the target is a company that has the ability and incentive to pursue R&D in the intended direction. The following example describes one of many such successful divestitures.

### Novartis–GSK (2015)

In 2015, Novartis agreed to acquire oncology assets from GSK. The FTC was concerned that the acquisition would harm competition in the development and future sale of inhibitors of the BRAF gene and MEK protein, medicines that inhibit molecules associated with the development of cancer. GSK was one of two firms with an approved BRAF-inhibitor drug, and the Commission believed that Novartis was the only other company with a BRAF-inhibitor in late-stage R&D. GSK also was the only firm with an approved MEK-inhibitor, and the Commission believed that Novartis was one of only a small number of firms with MEK-inhibitors in late-stage clinical development.[46]

To remedy its competition concerns, the Commission required Novartis to divest to Array BioPharma all rights and assets related to its BRAF- and MEK-inhibitor drugs in clinical trials.[47] The remedy appeared to restore innovation competition. Array's corporate documents indicate continued support for clinical trials of the drugs acquired from Novartis.[48]

The divestiture of the Novartis drugs to Array succeeded in restoring R&D incentives in part because Array is a company with a distinct focus on the discovery, development, and commercialization of small-molecule drugs for treating cancer. Thus, Array was an excellent fit to acquire the Novartis drugs. Moreover, the Novartis drugs were already in late-stage clinical trials and Array only had to choose whether to continue the existing trials.

## 5  Compulsory Licensing as a Merger Remedy

Compulsory licensing is an established tool to address competition concerns about mergers and acquisitions.[49] While compulsory licenses at noncompensatory rates may undermine innovation incentives for products that are easily copied, they can facilitate innovation and future price competition by reducing the cost of assembling necessary intellectual property rights, eliminating costly infringement litigation, enabling firms to supply complementary products, and neutralizing efforts by firms to use intellectual property rights to block new competition.

Chapter 6 reviews several important compulsory licensing decrees and academic studies of compulsory licensing that addressed competition concerns. With specific reference to mergers, Colleen Chien examines whether compulsory licensing orders in six merger cases caused the merged firm to reduce its patenting activity, and she finds no significant reduction in five of them.[50] Her research focuses on the effects of compulsory licensing on patenting by the licensor, while this section examines the effects of compulsory licensing obligations in merger remedies on patenting by licensors, licensees, and other relevant innovators. The experiences reported in this section are consistent with her findings and the studies reviewed in chapter 6: Compulsory licensing is often an effective mechanism to promote innovation and future price competition.

### Ciba-Geigy–Sandoz (1996)

Sandoz and Ciba-Geigy, through its ownership interest in Chiron, were early investigators into gene-based therapies. In 1996, the companies entered into an agreement to merge to form the pharmaceutical giant Novartis. According to the FTC, Ciba/Chiron and Sandoz were two of only a very small number of entities capable of commercially developing a broad range of gene therapy products. The Commission alleged that the merger, if not conditioned on remedies to restore competition, would likely harm innovation by eliminating or slowing the development of gene-therapy products and by reducing the incentive of the merged firm to license intellectual property rights or to collaborate with other companies.

To remedy the alleged competitive harm, the Commission conditioned its approval on patent licensing requirements and other restrictions. The final order obligated the merged company to grant nonexclusive rights at low royalties to patents covering proteins that regulate cell growth and to a broad patent covering the engineering of cells for gene therapies. In addition, the order required the merged company to grant a nonexclusive license covering certain gene therapy tools[51] to Rhône-Poulenc Rorer (RPR), and to either convert its exclusive license for use in gene therapy of the partial factor VIII gene (related to hemophilia) to a nonexclusive license or grant to RPR a sublicense to those gene therapy factor VIII rights, along with technical information and know-how if requested by a sublicensee. The Commission chose not to pursue divestitures of gene therapy assets because it felt that licenses were adequate to address the competition concerns, competitors possessed many of the hard assets necessary to compete, and asset divestitures would disrupt the parties' R&D efforts.[52]

The FTC believed that the stakes were extraordinarily high for gene therapies. The Commission endorsed projections that sales of gene therapy products would reach up to $45 billion by 2010 and that regulatory approvals for treatment of brain cancer tumors would occur by 2000. RPR was initially enthusiastic about gene therapies and planned to invest about $300 million to pursue new drugs.[53] Regrettably, gene therapies have failed to realize these expectations. A number of promising drugs turned out to be unsafe or ineffective. RPR merged with Hoechst Marion Roussel in 1999 to form Aventis, which was acquired by Sanofi in 2004. As of 2009, Sanofi had only one gene therapy drug in development,[54] which failed to show efficacy in a large phase III clinical trial.[55] The FDA did not authorize the use of any human gene therapy products until 2017, when it approved a Novartis treatment for acute lymphoblastic leukemia.[56]

It is not the fault of the FTC that the science of human gene therapy turned out to be much more challenging than the Commission had anticipated in 1996. The relevant question for the design of merger remedies is not whether gene therapies fulfilled early expectations but instead whether the compulsory licensing provisions in the Commission's order promoted industry R&D for gene-therapy drugs or possibly slowed R&D by limiting the ability of the merging firms to capitalize on their R&D expenditures.

Evidence does not indicate that the FTC's licensing order slowed R&D for gene therapies. As a measure of R&D effort, table 7.2 compares the number of successful patent applications in five-year windows before and after the merger that cite "gene therapy" in their description.[57] The first window ends one year after the merger to account for lags between R&D and the filing of relevant patents; that is, patents filed in 1996 and that ultimately issue are likely to reflect R&D performed in 1995, or possibly earlier.

The number of patents directed to gene therapies filed by the merged company in the five-year period following the merger was comparable to the number filed by the merging firms before the merger, while the total number of successful patent filings in this field approximately doubled. This suggests that the Commission's order had no adverse effect on patenting of gene therapy technologies and may have prevented the foreclosure of R&D activity for gene therapies that could have resulted from the merger.

Table 7.2
US patents that cite "gene therapy."

| Application Date | All US Patents | Ciba-Geigy, Chiron, Sandoz, or Novartis Patents |
| --- | --- | --- |
| January 1, 1993 to December 31, 1997 | 10,654 | 202 |
| January 1, 1998 to December 31, 2002 | 20,222 | 194 |

### Amgen–Immunex (2002)

Amgen proposed to acquire Immunex in December 2001. The transaction would combine two companies that were heavily invested in R&D for drugs developed with the use of recombinant DNA technologies. Among other concerns, the FTC identified possible harm to innovation for TNF inhibitors and IL-1 inhibitors. TNF inhibitors suppress responses to TNF, which is part of the body's response to inflammation. IL-1 inhibitors are molecules that bind to human interleukin-1, a class of proteins involved in the regulation of responses to inflammation. TNF and IL-1 inhibitors can be used to treat rheumatoid arthritis, among other illnesses. The Commission alleged that Amgen and Immunex were two of only a few companies with late-stage R&D projects for TNF inhibitors and IL-1 inhibitors.

In addition to other obligations, the Commission conditioned approval of the merger on the requirement that Amgen license certain patents related to TNF inhibitors to Serono, a Swiss biotechnology company with a soluble TNF inhibitor in clinical development. Amgen owned patents that the Commission alleged would block sales of Serono's TNF inhibitor in the US. To maintain competition for IL-1 inhibitors, it required the merged company to license Regeneron, the only other entity that it identified in late-stage R&D for an IL-1 inhibitor. This addressed a concern that Immunex had a patent portfolio that could block Regeneron's IL-1 Trap technology or create sufficient litigation costs to impede its development of a commercial product.

The consent order appeared to achieve its objectives. Table 7.3 compares the patenting activities of the merged Amgen and Immunex and Serono for patents that cite "TNF inhibition" in their description in the five-year periods before and after the acquisition and lists all US-issued patents in this category with application dates in these windows. As in table 7.2, the first window ends one year after the acquisition to capture patents that resulted from R&D effort before the acquisition. Amgen

Table 7.3
US patents that cite "TNF inhibition."

| Application Date | All US Patents | Amgen or Immunex Patents | Serono Patents |
|---|---|---|---|
| January 1, 1998 to December 31, 2002 | 5,530 | 165 | 18 |
| January 1, 2003 to December 31, 2007 | 6,475 | 198 | 42 |

Table 7.4
US patents that cite "Interleukin-1" or "IL-1."

| Application Date | All US Patents | Amgen or Immunex Patents | Regeneron Patents |
|---|---|---|---|
| January 1, 1998, to December 31, 2002 | 5,619 | 222 | 3 |
| January 1, 2003, to December 31, 2007 | 6,053 | 221 | 14 |

and Immunex, and the merged company, patented at about the same rate relative to all patents that reference TNF inhibition before and after the acquisition. Serono ramped up its patents from 18 in the five-year period before the acquisition to 42 in the five-year period following the acquisition. Significantly, Serono's annual reports during the five-year period following the merger of Amgen and Immunex describe its R&D efforts for a TNF inhibitor.[58]

Table 7.4 repeats this exercise for patents that cite "Interleukin-1" or "IL-1" in their description. Again, Amgen and Immunex patented at about the same rate relative to all patents before and after the acquisition, but Regeneron increased its patents from three to fourteen. Significantly, Regeneron's annual reports describe continuing R&D activities for its IL-Trap technology, which was a central objective of the Commission's consent order.

## Flow International Corp.–OMAX Corp. (2008)

Flow International develops, manufactures, and sells systems that use high-pressure waterjets that can be used in a wide range of cutting tasks. At the time of their proposed merger in 2008, Flow International and the OMAX Corp. were the largest and second-largest manufacturers of waterjet cutting systems, respectively.[59] OMAX also held two broad patents covering the controller that directs the cutting head. The FTC allowed the merger to proceed conditional on the grant of a

Table 7.5
US patents that cite "waterjet" and "cutting."

| Application Date | All US Patents | Flow International or OMAX Patents |
|---|---|---|
| January 1, 2005 to December 31, 2009 | 170 | 10 |
| January 1, 2010 to December 31, 2014 | 326 | 33 |

royalty-free license to the two OMAX patents to any firm that seeks a license.[60]

The licensing requirement appears to have achieved the objective of facilitating competition. One source identifies more than a dozen firms that offer waterjet cutting systems in addition to Flow International and OMAX, including five vendors that did not exist as recently as January 2014.[61]

A search for successful patent applications with the terms "waterjet" and "cutting" in their descriptions returned 170 patents in the pre-merger five-year window, of which 10 were invented by or assigned to Flow International or OMAX (table 7.5). In the post-merger five-year window, the total number of successful applications increased to 326, of which 33 were filed by the merged company. The numbers suggest that the compulsory license obligation had a beneficial effect on R&D effort for waterjet cutting systems.

### 3D Systems–DTM Corp. (2001)

In 2001, the DOJ challenged the proposed acquisition of DTM Corp. by 3D Systems. At the time, 3D and DTM were two of only a few suppliers of industrial rapid prototyping (RP) printing systems.[62] RP is a process by which a machine transforms a computer design into a three-dimensional model. The complaint alleged that the merger would lessen competition and harm innovation in the development, manufacture, and sale of industrial RP systems in the US.

The complaint also alleged that entry into RP was difficult, in part because 3D and DTM have extensive patent portfolios that relate to RP systems production.[63] The DOJ allowed the merger to proceed after the parties agreed to grant a license to develop, manufacture, and sell RP-related products, and to supply any necessary support or maintenance services, to at least one firm that currently manufactures industrial RP systems outside the US to facilitate its entry into the US market.[64]

This case straddles examples of targeted divestitures and compulsory licensing. The licensing requirement applied to an extensive portfolio of intellectual property, but like a targeted divestiture, it was limited to a single firm. In July 2002, the DOJ approved the Sony Corporation as the licensee for some of the merged company's patents and software copyrights, for use only in the field of stereolithography (a type of 3D printing) within North America, in exchange for a license fee of $900,000. The license does not apply to technology that 3D-DTM may develop in the future.[65]

The divestiture order succeeded in facilitating new competition[66] and did not appear to adversely affect patenting for rapid prototyping printing systems.[67] Furthermore, 3D Systems accelerated R&D spending following the order and its merger with DTM.[68] The company spent an average of $9.2 million per year on R&D in the three-year period (1998–2001) preceding the merger and spent an average of $11.6 million per year in the three-year period (2002–2005) following the merger.

Several other firms crossed the patent barrier and became significant suppliers of RP systems since the time of the 3D-DTM merger, and sales of RP systems by 3D Systems and Sony in the US lagged sales by many other companies. One source showed 3D Systems in fourth place among vendors of industrial 3D printing systems as of April 2018, lagging behind Stratasys, EOS, and GE Additive.[69] In hindsight, the remedy agreed to by the DOJ and the merger parties was not unproductive, but it was arguably unnecessary.

## 6  Innovation Enforcement at the European Commission

The EC did not allege harm for innovation in a merger case until the 2000s.[70] In 1994, the Commission reviewed a proposal by Pasteur-Mérieux and Merck to establish a joint venture to manufacture and distribute vaccines in Europe. The Commission recognized overlaps of vaccine projects in the companies' R&D pipelines but chose not to challenge the joint venture, citing its technological benefits.[71] In 1995, the EC and the FTC investigated the proposed merger of the pharmaceutical companies Glaxo and Wellcome that formed Glaxo Wellcome (GW). Both agencies identified a potential loss of product-to-project competition for noninjectable drugs to combat migraine attacks. The EC concluded that the merger did not threaten innovation because there was sufficient R&D competition from other large pharmaceutical companies.[72] The FTC disagreed; it conditioned approval of the merger on a requirement that Glaxo divest Wellcome's worldwide R&D assets for these drugs to replace the competition lost from the merger.[73] This is an example of a successful divestiture for R&D. Zeneca acquired the assets. Its successor company, AstraZeneca, won FDA approval for a new migraine drug, zolmitriptan, in 1997.[74]

The EC and the FTC again diverged in their evaluations of innovation competition when GW and SmithKline Beecham merged in 2001 to form GSK. As previously discussed in this chapter, the FTC identified a loss of innovation competition for topoisomerase I inhibitor drugs used to treat certain types of cancers and for a herpes vaccine, although its required divestitures had questionable benefits. The EC recognized that the merging parties had overlapping projects in their R&D pipelines, but concluded that the merger was unlikely to have a significant adverse effect on R&D efforts. With respect to prophylactic herpes vaccines, the EC regarded the research as too uncertain and, if research proved to be successful, future products would be too distant to predict competitive effects.[75] The EC also noted the large number of active competitors for new cancer drugs.[76]

The EC subsequently pursued innovation competition concerns with more vigor in its merger enforcement. Most of the Commission's actions paralleled the outcomes of merger evaluations by the US FTC and DOJ. The agencies reached similar conclusions regarding possible harm to innovation in their evaluations of the merger of AstraZeneca and Novartis that formed the agrochemical giant Syngenta;[77] the acquisition of Aventis Crop Science by Bayer;[78] two proposed mergers of manufacturers of personal computer hard disk drives;[79] the acquisition of the medical device manufacturer Covidien by Medtronic;[80] and several other transactions.

The EC and US antitrust authorities have reached different conclusions in some recent evaluations of innovation effects from mergers. In its review of Pfizer's acquisition of Hospira in 2015, the EC cited harm to innovation to justify a requirement that Pfizer should divest a drug that it had under development that would potentially compete with an existing drug comarketed by Hospira.[81] In contrast, the FTC reviewed the transaction but took no action regarding a lessening of innovation competition.[82]

The drug sold by Hospira was a biosimilar[83] for the biologic drug Infliximab (trade name Remicade), an expensive drug that is prescribed to treat a range of autoimmune conditions. Hospira marketed its biosimilar under the trade name Inflecta. The EC concluded that Pfizer's acquisition of Hospira's biosimilar would cause Pfizer to either delay or discontinue its own biosimilar for Infliximab that was in the company's R&D pipeline, leading to a "lessening of innovation competition."[84]

The EC was correct to require Pfizer to divest its biosimilar project, but for the wrong reason. The Hospira acquisition did not cause Pfizer to delay or discontinue its pipeline project. Novartis acquired and developed the project for sale in the European Economic Area (EEA). The EC's order left Pfizer with the rights to develop and market Infliximab biosimilars outside the EEA. Pfizer pursued FDA approval for its second Infliximab biosimilar, which it branded with the trade name Ixifi. However, to date Pfizer has elected not to sell Ixifi in the US, where it would compete against its Inflectra biosimilar, as well as a third biosimilar from another company and Remicade.[85] The acquisition lessened future price competition, which the EC's divestiture remedied in the EEA and which the FTC might have remedied in the US if it had also required Pfizer to divest its biosimilar project to an appropriate acquirer. The case is unusual in that Pfizer could elect to sell Ixifi in regions where it did not sell Inflectra. It is possible that Pfizer would have delayed or discontinued Ixifi if it did not have this option.

The EC and the US FTC and DOJ often make allegations of innovation harm that supplement traditional antitrust allegations of harm to competition for existing product markets. An example is the discussion of innovation harm in the EC's evaluation of the acquisition by General Electric of Alstom's heavy-duty gas turbine business.[86] The EC emphasized the competitive significance of Alstom as an independent innovator in a concentrated industry, but the loss of price competition from the transaction likely was sufficient justification to challenge the transaction. Nonetheless, it is useful to include relevant innovation claims in litigated cases because they allow courts to become more familiar with these types of allegations.

In other cases, innovation allegations are central to enforcement decisions and shape remedies to address the loss of competition. In most of these cases, the agencies addressed harms from the loss of identified product-to-project or project-to-project innovation competition. These cases are toward the base of the "innovation enforcement pyramid" (figure 5.5 in chapter 5). The few cases that alleged harm from the concentration of overlapping R&D capabilities have mostly addressed industries in which only two or three firms have the requisite capabilities, such as in the challenge by the DOJ to the GM-ZF merger.

The EC's challenge to the merger of the agrochemical businesses of Dow and DuPont marks an important departure from this pattern.[87] At the time of the proposed merger, valued at approximately $130 billion, Dow and DuPont were two of a handful of large integrated R&D firms that developed, manufactured, and sold crop protection chemicals. The

Antitrust Division of the DOJ (along with several states) investigated the proposed merger and identified concerns about the elimination of head-to-head competition in markets for broadleaf herbicides for winter wheat and insecticides for chewing pests. Although the complaint also mentions that the merger would eliminate innovation rivalry by two of the leading developers of new crop protection chemicals, the parties accepted a consent decree that only required DuPont to divest one of its herbicide products and one of its insecticides, along with the assets required to develop, manufacture, and sell these products.[88]

The EC also investigated the merger and expressed concerns that included, but went much deeper than, the isolated overlaps identified by the US Antitrust Division. In a merger decision with more than 600 pages and almost 300 pages of annexes, most of which addresses crop protection chemicals, the Commission concluded that the merger would remove the parties' incentives to pursue ongoing parallel innovation efforts for a number of important herbicides, insecticides, and fungicides.[89] As a condition to approve the merger, the EC required DuPont to divest a significant part of its existing pesticide business, including almost all of its global R&D organization for agricultural pesticides.[90]

The EC supported its decision with references to the economic theory of unilateral innovation harm. It concluded that prior to the merger, each party could contest the other party's business by investing more in R&D, and that discoveries are protected with strong patent rights, which reduce merger benefits from the internalization of spillovers. According to this theory, the merger would eliminate a profitable contestability incentive, giving rise to a unilateral effect to reduce investment in R&D. The EC also identified the risk that R&D projects undertaken by one of the merging parties would replace profits from sales of products by the other party; that is, concerns about the loss of product-to-project competition. In addition, it reviewed the history of agrochemical mergers and innovation and corporate documents, from which it inferred evidence that the merger would harm innovation.

The novel element in the EC's allegations is the application of a unilateral effects theory for early-stage R&D incentives in an industry with more than a few potential innovators. Many firms and public institutions invest in agricultural research. Although upstream R&D for agricultural pesticides is not highly concentrated, the EC limited the universe of innovators to companies that invest in all stages of the value chain (i.e., discovery, development, mixture/formulation, and commercialization). According to the EC, only Syngenta, Bayer, BASF, Dow, and DuPont fit this description. In addition to excluding many small firms and R&D laboratories, this definition excluded FMC because it allegedly had exited R&D for new pesticides, and Monsanto because its R&D was almost exclusively focused on seeds.

Furthermore, the EC argued that R&D activities are clustered in "innovation spaces," which are groupings of crop/pest combinations in which there are often only four or fewer actual or potential innovation competitors.[91] These clusters excluded Japanese companies that research and develop pesticides for rice and other crops in their geographic area and that only occasionally have application in European agriculture.

Patent data assembled by the EC illustrates the implications of its innovation space approach. Limiting the R&D universe to the five integrated companies and the top 25 percent of patents measured by number of external citations, the EC concluded that the merger would increase the patent Herfindahl-Hirschman Index (HHI) by about 800–900 for herbicides and by 1,100–1,200 for insecticides, resulting in post-transaction HHIs of 3,000–3,500.[92] These numbers indicate concentration effects that are significant, and arguably even higher in narrow innovation spaces. If the analysis broadens to include patents filed by Japanese firms (while still excluding many other firms and institutions involved in agrochemical R&D), the changes in the HHI from the merger fall to 400–600, with posttransaction HHIs of only 1,500–2,500. These numbers indicate far lower concentration effects.

Economic evidence for harm to innovation from the Dow-DuPont merger is inconclusive given the multiplicity of firms engaged in R&D for agricultural pesticides. Of course, innovation outcomes cannot be predicted with certainty and there are errors of under-enforcement as well as over-enforcement. Furthermore, it is possible that the Dow-DuPont divestiture will prevent prices increases for future agricultural chemical products. The number of firms with the ability to test, manufacture, and distribute these chemicals is much smaller than the number of firms that can invest in R&D to discover promising new molecules. Curiously, although the EC mentions concerns about the loss of diversity and future product competition,[93] its decision places more emphasis on its prediction of a loss of innovation competition than on the loss of future price competition.

## 7  Lessons for Merger Policy for Innovation

The case study evidence in this chapter generally supports the positive role for antitrust enforcement in preserving innovation incentives. The examples of unconditional merger challenges in this chapter do not identify harm to innovation from abandoned transactions. The firms in these examples continued to innovate as independent entities, and there is no evidence that the merger challenges caused the firms to ratchet down their R&D activities.

The case study evidence also identifies risks associated with R&D divestitures that are negotiated as remedies to approve a merger. Some R&D divestitures surveyed in this chapter replaced lost competition and preserved incentives to invest in R&D, while others were followed by abandoned R&D projects. The successful examples typically involved specific projects in a late stage of R&D that were acquired by firms with a demonstrated ability and incentive to promote the acquired projects. It is possible that some firms that acquired early-stage R&D projects viewed these projects as options to acquire human capital and intellectual property at a bargain-basement price.

Does it follow from these examples that the antitrust agencies should stiffen their backs and block more proposed mergers rather than allow targeted divestitures? While there is ample cause for concern that some mergers will harm innovation and future price competition, unconditional challenges may not erase these harms. If merging firms have a large portfolio of products and R&D activities, and if they expect a challenge resulting from overlaps in one or two areas, they could fix the transaction by divesting the problematic assets before they face antitrust review or during the review process. If an antitrust agency cannot negotiate a divestiture that restores alleged lost innovation competition, it is not likely that a divestment negotiated by the merging parties would be more successful.

The cases reviewed in this chapter and the studies reported in chapter 6 show that compulsory licensing obligations generally have not had adverse effects on industry patenting activity or R&D investment and appeared to have promoted R&D activity and competition in several instances. Of course, antitrust enforcement must respect the risk that overuse of compulsory licensing could have disincentives for R&D. However, contrary to claims that compulsory licensing threatens innovation,[94] there is no evidence from merger consent decrees or other examples of compulsory licensing that compulsory licensing obligations, when selectively applied to address competition concerns, have generally negative effects on incentives for R&D that outweigh their procompetitive benefits.

# 8 "We Are Going to Cut Off Their Air Supply": Microsoft and Innovation Harm from Exclusionary Conduct

> We decide this case against a backdrop of significant debate amongst academics and practitioners over the extent to which "old economy" §2 monopolization doctrines should apply to firms competing in dynamic technological markets characterized by network effects.
> —*US v. Microsoft*, Court of Appeals (2001)

## 1 Introduction

More than two decades have passed since the US Department of Justice (DOJ), several states, and the European Commission (EC) accused Microsoft of violating antitrust laws.[1] The allegations stirred controversy, with observers expressing concerns that ranged from warnings that antitrust enforcement would chill innovation in the personal computer (PC) industry to expressions of outrage over missed opportunities for stronger remedies.[2] The industry has changed dramatically since these cases were resolved. Nonetheless, the Microsoft antitrust cases still hold valuable lessons for how courts should evaluate conduct in high-tech markets.

In the 1990s, Microsoft held a commanding position as the supplier of the world's most popular PC operating system, Microsoft Windows. Windows was installed on more than 80 percent of IBM-compatible PCs, at the time by far the world's most popular type of PC.[3] Microsoft's grip on the PC market seemed secure. The operating system was a technical achievement protected by a web of intellectual property rights. Yet the real key to Microsoft's security was not its technical complexity or intellectual property, but rather the thousands of applications written to operate with Windows.

Personal computer operating systems are platforms with strong cross-platform network effects. Computer users value the number and quality of applications that run on the operating system, and application developers value the number of operating system users. Motivating developers to write applications that would run on a new operating system is a barrier to entry that new operating system entrants must cross to become viable competitors.

IBM confronted the applications barrier after it spent more than $1 billion to develop, test, and market a PC operating system to compete with Windows.[4] IBM created the first mass-market PC, but it did not have proprietary software or hardware that would enable the company to fully monetize its technologies. Frustrated by its inability to capitalize on the burgeoning PC industry, IBM launched an alternative PC operating system called OS/2 Warp in 1994.[5] IBM ultimately abandoned OS/2, not because it was technologically inferior to Windows, but because it failed to attract a sufficient number of applications to make computer users switch from Windows.[6]

Despite Microsoft's success in suppressing challengers, in 1995 Microsoft's chairman and chief executive officer, Bill Gates, was nervous about an emerging threat to its empire. Microsoft was behind the curve on a new technology, the internet, which Gates correctly characterized as "the most important single development to come along since the IBM PC was introduced in 1981."[7] Internet users were connecting to the World Wide Web with a new program, the Netscape Navigator internet browser. Gates was nervous not merely because Microsoft was late to the internet party, but also because he was concerned that the browser was a Trojan horse that would break the applications barrier to entry by creating an alternative platform for software development.

Navigator is a type of software called middleware, which occupies a space between the operating system and applications. Like operating systems, middleware facilitates application software development by exposing application programming interfaces (APIs), which are routines that perform certain widely used functions. Unlike operating systems, the Netscape APIs were not specific to a particular operating system. Netscape employed a new programming language, Java, which allows applications to run on different operating systems.[8] Gates warned his colleagues within Microsoft that Netscape was "pursuing a multi-platform strategy where they move the key API into the client to commoditize the underlying operating system."[9] Netscape's cofounder, Marc Andreessen, fueled Gates's anxiety when he predicted that the browser would become a meta-operating system and reduce the Microsoft operating system to "an unimportant collection of slightly buggy device drivers."[10]

Microsoft should have welcomed Netscape, if not for the threat that it posed to the operating system. Internet browsing technologies are complements to other services provided by PCs, and in that respect they add value to Microsoft's operating system by giving consumers access to the World Wide Web. The concern that rattled Gates was the risk that Netscape would allow rival operating systems to benefit from the same network effects that fueled demand for Microsoft's operating system (namely, the benefits that users obtain from the ability to use a wide range of applications). Network effects had caused the PC market to tip to the Windows platform by the mid-1990s.[11] Gates understood that middleware such as Netscape, which promoted the use of the platform-independent Java programming language, could level the PC market and fracture the Windows monopoly.

Microsoft's initial response to the Netscape threat was an attempt to exclude Netscape from supplying browsers that operated with Windows operating systems. Microsoft executives met with top Netscape personnel in May 1995 and suggested that the companies divide the browser market, with Microsoft becoming the sole browser supplier for Windows in return for Microsoft refraining from competing with Netscape for browsers that work with other operating systems. Netscape did not accept Microsoft's offer.[12] Had Netscape accepted, antitrust authorities could have challenged the arrangement as a collusive market division.

Gates had another solution to plug the vulnerability in the applications barrier wall. Microsoft had its own browser, Internet Explorer (IE), but Netscape had gotten a head start; in 1996, IE accounted for only about 5 percent of internet usage. Netscape sold its browser at that time as a separate product, with a retail price of $49. Microsoft originally sold IE as a separate product as well, for use with Windows 95. But Gates could offer IE for free, as a bundle included with Windows 95, and Microsoft subsequently integrated IE into newer versions of its operating system. Windows was Microsoft's ace in the hole. Nearly every purchaser of an IBM-compatible PC wanted the computer to be equipped with the Windows operating system. By bundling IE with the operating system, consumers didn't pay extra for the Microsoft browser, but if they wanted Navigator as an add-on, they would have to pay an additional $49 or so. A senior Microsoft executive was quoted as saying, "We are going to cut off their air supply."[13] The Windows/IE bundle robbed Netscape of the demand (air) to pay for Navigator.

Microsoft employed numerous other tactics to suppress adoption of the Navigator browser and advantage IE in what came to be known as the first "browser war."[14] Microsoft's efforts to suppress Netscape, along with improvements to IE, allowed the company to increase its IBM-compatible PC browser share from about 5 percent in 1996 to more than 50 percent by 1998.[15]

Section 2 describes the case filed by the DOJ and several states which alleged that Microsoft engaged in unlawful conduct to *maintain* its Windows monopoly. Section 3 describes a case brought by the EC, which raised a different concern: unlawful conduct by Microsoft to *extend* its monopoly in PC operating systems into the market for work-group server operating systems.[16] While the DOJ and the states focused on efforts by Microsoft to prevent the adoption of platform-independent middleware, the EC focused on features of Windows that made it incompatible with non-Microsoft server software. The EC also pursued bundling allegations for media players and browsers.[17]

Section 4 examines the US and European antitrust allegations from an economic perspective. The antitrust authorities in both jurisdictions concluded that many aspects of Microsoft's conduct harmed innovation. Nonetheless, for the most part they applied conventional antitrust jurisprudence to evaluate whether Microsoft's conduct violated their antitrust laws. An exception is the refusal by the court of appeals in the US case to hold Microsoft liable for tying IE to the Windows operating system, which the court instead remanded for further analysis by the district court to account for efficiency justifications from selling software bundles or integrated software products.

A central premise in the US case—and to a lesser extent the European case—is the potential for browsers and other middleware to eliminate Microsoft's dominance of PC operating systems. That potential has not been realized. Nevertheless, the cases prevented Microsoft from extinguishing rival browsers and encouraged Microsoft to open its software products to interoperate with other products. The cases also affirmed the legal precedent that it is unlawful for dominant firms to extinguish nascent competitive threats without requiring proof that the threats are significant and likely to be realized.

The US and European plaintiffs settled their allegations without further litigation or harsh punishments such as a breakup. Section 5 asks whether the settlement was a lost opportunity. Although many criticized the settlement for being too soft on Microsoft, the terms of the settlement ended Microsoft's exclusionary conduct, facilitated competition for browsers and other middleware, and encouraged Microsoft to support interoperable products. Section 6 concludes with lessons gained from the US case that are applicable to other competition issues in the high-technology economy.

## 2 US v. Microsoft

The DOJ sued Microsoft on May 18, 1998, for monopolizing the markets for PC operating systems and browsers in violation of Sections 1 and 2 of the Sherman Act. Nineteen states and the District of Columbia joined the suit. The complaint alleged that Microsoft excluded competition in browsers and PC operating systems by "tying" IE to Windows. The complaint alleged a tie even when Microsoft integrated IE into Windows because Microsoft refused to sell IE as a separate product (a "technological tie"). In addition, the complaint alleged that Microsoft excluded competition by entering into agreements with PC original equipment manufacturers (OEMs), internet service providers (ISPs), internet content providers (ICPs), and independent software vendors (ISVs) that favored IE and disadvantaged Navigator.

By 1999, when the case went to trial before Judge Penfield Jackson in the District Court of the District of Columbia, the DOJ and the plaintiff states had brought additional allegations that were not in the original complaint.[18] These allegations buttressed the plaintiffs' case that Microsoft sought to prevent the emergence of middleware that would enable applications that were not dependent on the Windows platform. An allegation aimed directly at the ability of the browser to break the applications barrier to entry addressed Microsoft's actions to subvert the adoption of Java to enable interoperable software. Microsoft licensed Java from Sun Microsystems and purported to promote Sun's platform-independent Java technologies. The government and the plaintiff states alleged that: Microsoft modified the Java programming language in a way that made it incompatible with Sun's implementation; entered into contracts with major ISVs to promote Microsoft's proprietary version of Java rather than any Sun-compliant version; did not adequately inform ISVs that its version of Java did not conform to the Sun implementation; and coerced Intel to stop working with Sun Microsystems to improve the Java technologies.[19]

The complaint specifically alleged harm to innovation from Microsoft's conduct. The harm included the following:[20]

• Impairing the incentive and ability of Microsoft's competitors to undertake research and development (R&D) by limiting their rewards from any resulting innovation
• Impairing the ability of Microsoft's competitors and potential competitors to obtain financing for R&D
• Inhibiting Microsoft's competitors that succeed in developing promising innovations from effectively marketing their improved products to customers
• Reducing the incentive and ability of OEMs to innovate and differentiate their products in ways that would appeal to customers
• Reducing competition and the spur to innovation by Microsoft and others that only competition can provide

Judge Jackson concluded that Microsoft engaged in a campaign of exclusionary conduct that "succeeded in preventing—for several years, and perhaps permanently—Navigator and Java from fulfilling their potential to open the market for Intel-compatible PC operating systems to competition on the merits"[21] and did not advance any legitimate business justifications for its conduct.[22] The threats that middleware and platform-independent Java programming tools posed to Microsoft were central to the court's findings, although the court acknowledged that "these middleware technologies have a long way to go before they might imperil the applications barrier to entry."[23]

The court held Microsoft liable for monopolizing the market for the worldwide licensing of Intel-compatible PC operating systems, and for attempting to monopolize the market for internet browsers, both in violation of Section 2 of the Sherman Act. In addition, it found Microsoft guilty of unlawful tying, in violation of Section 1 of the Sherman Act, by requiring consumers to take IE as a condition of obtaining Windows.[24]

Judge Jackson concluded:[25]

While the evidence does not prove that [new competitors] would have succeeded absent Microsoft's actions, it does reveal that Microsoft placed an oppressive thumb on the scale of competitive fortune, thereby effectively guaranteeing its continued dominance in the relevant market. More broadly, Microsoft's anticompetitive actions trammeled the competitive process through which the computer software industry generally stimulates innovation and conduces to the optimum benefit of consumers.

The court accepted the plaintiffs' proposed remedy. In addition to a series of temporary conduct restrictions, the proposed remedy called for the structural separation of Microsoft into two independent corporations, with one continuing Microsoft's operating systems business and the other undertaking the balance of the company's operations.[26]

Microsoft appealed. The court of appeals sustained the district court finding that Microsoft monopolized the market for the licensing of Intel-compatible PC operating systems, but it rejected the finding that Microsoft attempted to monopolize the market for internet browsers, largely because the plaintiffs did not establish a separate market for internet browsers or show why such a market, if it existed, could be monopolized. The court also refused to affirm Judge Jackson's finding that Microsoft engaged in a per se illegal tie when it bundled the sale of IE with Windows 95, and later when Microsoft integrated IE into Windows 98.[27]

In rejecting the per se rule for the tying of software products, the court of appeals noted, "We do not have enough empirical evidence regarding the effect of Microsoft's practice on the amount of consumer surplus created or consumer choice foreclosed by the integration of added functionality into platform software to exercise sensible judgment regarding that entire class of behavior."[28] The court remanded the tying issue for a rule of reason analysis that would have accounted for efficiency justifications from selling software bundles or integrated software products, but the parties ultimately agreed to a settlement that preempted this evaluation.

A key issue in the case was whether Microsoft undermined industry adoption of a platform-independent Java programming language by supporting a proprietary version of Java and encouraging software developers to use its version rather than the platform-independent version promoted by Sun Microsystems. Judge Jackson held that these actions were part of an overall scheme to monopolize the markets for Intel-compatible operating systems and internet browsers. The court of appeals agreed that Microsoft's actions with regard to Java were exclusionary, but it held that "a monopolist does not violate the antitrust laws simply by developing a product that is incompatible with those of its rivals" and that Microsoft had sufficient legitimate reasons to promote its modified version of Java to escape a finding of antitrust liability.[29]

On the critical issue of the structural remedy, the court of appeals ruled that the district court failed to hold a hearing on the likely consequences from the ordered relief and failed to provide an adequate explanation for the order. Accordingly, the court of appeals vacated the court's remedy order in its entirety and remanded the case with instructions to conduct a remedies-specific evidentiary hearing and a new determination in light of the appellate court's more limited findings of liability.[30]

Judge Jackson had several interactions with the press during the Microsoft trial, in which he lambasted Microsoft's conduct. At one point, he likened the structural remedy to training a mule by hitting it with a two-by-four to get its attention. The court of appeals took issue with his comments to the press and remanded the case to a different district court trial judge for the hearing on remedies.

A full hearing on remedies never occurred. The Antitrust Division of the DOJ and nine state plaintiffs agreed to a settlement that abandoned the structural remedy in favor of imposing a number of behavioral conditions. Nine other plaintiff states returned to the courthouse to litigate a more severe remedy, but they agreed to a Final Judgment (Settlement) in November 2002 that differed only slightly from the one proposed by the Antitrust Division and the nine settling states.[31] The Final Judgment included requirements that Microsoft not punish OEMs or ISVs for supporting competitive products. Microsoft agreed to give OEMs flexibility in the way that they display icons, shortcuts, and menu options, including the promotion of non-Microsoft middleware. Furthermore, the Final Judgment compelled Microsoft to disclose APIs, communication protocols, and the related documentation necessary for middleware to communicate with the operating system and with a Microsoft server operating system product; and it also obligated Microsoft to license any intellectual property that was necessary to supply the relevant interoperability information at reasonable and nondiscriminatory terms.[32]

## 3   The European Commission Enters the Fray

In December 1998, the EC opened an investigation into allegations that Microsoft harmed competition by (1) unlawfully tying its Windows Media Player (WMP) to the Windows operating system and (2) designing Windows 2000 to make it incompatible with rival workgroup server operating systems. The EC ultimately held that both allegations were an abuse of dominance, in violation of European antitrust law.[33]

The tying claim paralleled the browser-tying allegation in the US case, but it had a different outcome. In the US case, the court of appeals acknowledged potential efficiency justifications for the tying of software products and remanded the district court's decision for further review under the rule of reason. The EC considered, but rejected,

Microsoft's tying defenses. The EC did not credit Microsoft's claim that the WMP was an integral component of Windows, or that the tie lowered consumer transaction costs by providing the WMP as a convenient default option. Furthermore, the EC did not consider whether other media players could compete without charging fees (e.g., by collecting revenues for advertising or content).

The EC ruled, "The tying of WMP... shields Microsoft from effective competition from potentially more efficient media player vendors, which could challenge its position, thus reducing the talent and capital invested in innovation in respect of media players." In addition to a fine for abuse of dominance, the EC settled the WMP allegation by requiring Microsoft to sell a version of Windows without the media player, which it did at the same price. Not surprisingly, there were few sales of the bare-bones version of Windows.[34]

In response to the interoperability allegation, the EC concluded that Microsoft's failure to provide interoperability information imposed a competitive disadvantage on rivals in the market for workgroup servers, in violation of European antitrust law.[35] The EC rejected Microsoft's claim that some of the information was protected by intellectual property and that Microsoft's decision not to license its intellectual property was an objective defense to the interoperability allegation.[36] In 2004, the EC ordered Microsoft to offer to potential rivals, with no expiration date, "complete and accurate specifications for the protocols used by Windows work group servers in order to provide file, print, and group and user administration services to Windows work group networks"[37] and to license them at reasonable and nondiscriminatory terms.[38]

In 2008, the EC opened new Microsoft investigations into the tying of the IE browser to Windows and the disclosure of interoperability information.[39] The latter ended in response to voluntary commitments by Microsoft to supply information to improve interoperability between third-party products and several Microsoft products, including Windows, Windows Server, Microsoft Office, Microsoft Exchange, and Microsoft SharePoint.[40] The browser tying investigation was settled after Microsoft agreed to pay fines and include a "ballot box" in its start-up display that allowed users to install different browser products and choose a default browser.[41] That obligation expired at the end of 2014,[42] and aptly so because IE's share of browser usage was half that of the Google Chrome browser by that date.[43]

The interoperability allegations and resulting settlements have special significance because they addressed conduct that is difficult to reach under US antitrust law but nonetheless affects the ability of firms to compete in a critical sector of the information economy. US antitrust law imposes few obligations on firms to license intellectual property or otherwise assist their competitors. In *Verizon v. Trinko*, the plaintiff alleged that Verizon had discriminated against rivals that sought to compete in local telephone markets by delaying or impeding their connections to Verizon's lines. The US Supreme Court held that Verizon's failure to accommodate rivals was not unlawful exclusionary conduct. In its 2004 opinion, the Court said, "Firms may acquire monopoly power by establishing an infrastructure that renders them uniquely suited to serve their customers. Compelling such firms to share the source of their advantage is in some tension with the underlying purpose of antitrust law, since it may lessen the incentive for the monopolist, the rival, or both to invest in those economically beneficial facilities."[44]

European antitrust law is more accommodating to allegations that dominant firms have a duty to assist their rivals.[45] As precedent for its decision that Microsoft harmed competition and innovation by withholding interoperability information, the EC cited a commitment by IBM in the early 1980s to disclose sufficient interface information to competitors in the European Community to enable them to attach their hardware and software products to IBM's System/370 mainframe.[46] The EC also cited a 1991 European Software Directive, which, with regard to limitations on the rights of copyright holders, states the objective "to make it possible to connect all components of a computer system, including those of different manufacturers, so that they can work together."[47]

There are other cases in which the EC asserted a duty for dominant firms to assist rivals. In a 1993 decision, the EC ruled, "An undertaking in a dominant position may not discriminate in favour of its own activities in a related market."[48] In two prominent cases that preceded the Microsoft decision, the EC ruled that copyright protection did not immunize owners from an obligation to share information with their rivals.[49]

## 4   An Economic Assessment of the US and European Antitrust Litigation against Microsoft

The Microsoft litigation was one of the most important episodes of antitrust enforcement action in the high-technology economy. This section offers an economic perspective on whether the litigation effectively addressed key competition issues, with a focus on innovation.

### Did the US courts enforce the antitrust laws to promote innovation?

The complaint filed by the DOJ and several states highlighted concerns about the effects of Microsoft's conduct on innovation in the PC industry. Judge Jackson's Findings of Fact mentions "innovation" no less than thirty times. The court of appeals introduced its opinion with the statement, "We decide this case against a backdrop of significant debate amongst academics and practitioners over the extent to which 'old economy' §2 monopolization doctrines should apply to firms competing in dynamic technological markets characterized by network effects" and noted that "competition in such industries is 'for the field' rather than 'within the field'." The court also cited Joseph Schumpeter to support the proposition that monopoly may be a temporary phenomenon in these types of markets.[50] In reaching its opinion, however, the court of appeals did little to assess whether Microsoft's conduct promoted or harmed innovation.

The court of appeals applied a rule of reason analysis to three elements related to the design of IE and Windows: (1) Microsoft's elimination of IE from the Add/Remove Programs utility in Windows 98, (2) the comingling of software code for Windows and IE, and (3) Microsoft's refusal to allow OEMs to uninstall IE or remove it from the Windows desktop. The court sustained findings that the first two elements had anticompetitive features and that Microsoft did not proffer procompetitive justifications. For the third element, the court recognized Microsoft's procompetitive justifications and held that the plaintiffs did not meet their burden to demonstrate anticompetitive effects.[51] The court avoided the difficult task of balancing anticompetitive effects and procompetitive justifications from Microsoft's product design decisions. In all three instances of design-related conduct that the court examined closely, it found either an anticompetitive effect or a procompetitive justification, but not both.[52]

A different element of the court's decision affirmed a valuable precedent for the evaluation of innovation and future competition. The court sustained the verdict that Microsoft unlawfully maintained its monopoly in IBM-compatible PC operating systems by eliminating the *nascent* threat posed by Netscape and other middleware. The court did not require evidence of harm to actual competition, nor did the court require proof that potential harm to competition was likely.

Should the courts have gone further to assess innovation concerns? There was no dispute that Microsoft was a highly innovative company when the DOJ and the states filed suit. From 1994 to 1996, Microsoft invested almost 15 percent of its revenue in R&D, which was far above the average R&D intensity in manufacturing, and spent millions of dollars to develop and improve IE. Microsoft was compelled to innovate in order to sell operating system software to new customers and upgrades to its installed base of customers that already owned a computer with Windows.

The relevant question for antitrust enforcement is not whether Microsoft was an innovative company, but whether Microsoft's challenged practices hindered innovation. Chapter 3 explains that a firm with monopoly power might invest heavily in R&D to preempt new competition. As a monopolist, Microsoft had more to gain by preventing competition than rivals could expect to achieve by competing head-to-head against Microsoft. But Microsoft did not preempt competition by innovating more than its rivals. Instead, Microsoft entered into contractual arrangements to limit Netscape's access to key distribution channels and to impede the adoption of platform-neutral software. This is exclusionary contracting, not preemptive innovation.

Antitrust enforcement for exclusionary contracts can alter the trade-offs among incentives to create a new product, incentives for subsequent innovators, and the consumer benefits from the innovations. Weak antitrust rules that allow a firm to lawfully exclude rivals increase the incentive to invent by increasing the profit from the innovation. But weak antitrust rules also allow an incumbent to exclude subsequent innovators and charge high prices.[53]

Finding the optimal balance between rewards for initial and subsequent innovation can be a challenging task in the real world of antitrust enforcement.[54] Nevertheless, it is clear that Microsoft profited handsomely as a successful incumbent, and its conduct erected barriers to subsequent innovators. It was far more important for antitrust enforcement to increase the opportunity for firms to compete within the PC field rather than allow Microsoft to profit by shielding itself from competition. There is no evidence that Microsoft's exclusionary conduct was justified because it increased the company's incentive to innovate.

There is an argument that antitrust enforcement against Microsoft was futile because network effects in the PC industry would make a dominant supplier an inevitable market outcome. In fact, experience shows that rivals can coexist, but this argument would be flawed even if the market were winner-take-all. Competition *for* the market is a spur to innovate notwithstanding that powerful network effects may shield the successful innovator from further competition *in* the market. Absent

Microsoft's exclusionary conduct, rivals have large incremental payoffs from R&D, which can make the difference between a large return or no return at all if another rival is first to capture the market. This larger incremental return from R&D is a spur for innovation in a competitive market.

The district court held that Microsoft's agreements contributed to monopolization, in violation of Section 2 of the Sherman Act, but held that they did not violate Section 1 of the Sherman Act because they did not substantially foreclose competition. Considering the role of network effects in the PC industry, an economic case can be made that the agreements also should have been held unlawful under Section 1 because, if the agreements were allowed to persist, their exclusionary effects would have been sufficient to tip the market to IE and prevent competition from Netscape and other platform-independent middleware.

Microsoft develops and operates Windows as a platform that coordinates applications developers, device manufacturers, and computer users. Neither Judge Jackson nor the court of appeals specifically analyzed Microsoft's conduct as a platform, perhaps because platforms were not part of the vernacular at that time. A two-sided analysis may have affected a finding of liability related to the integration of Windows and IE (which the appellate court remanded for further analysis). The integration of IE with Windows made APIs available to software developers to implement internet functionality, such as code that executes Hypertext Markup Language (HTML) files. This is a benefit that accrues to the application side of the Windows platform at the possible expense of computer users and OEMs that would prefer IE to be less integrated with Windows. It is unlikely, however, that a two-sided or multisided analysis would have reversed the finding that Microsoft's exclusionary contracts violated the Sherman Act. There is no evidence that Microsoft's restrictive agreements promoted innovation on either side of the Windows platform.

At a general level, economic theory regarding the effects of competition on innovation incentives supports the allegations brought by the DOJ and several states against Microsoft and the findings of antitrust liability.

### The European Commission case: Did interoperability obligations promote innovation?

The EC stressed the procompetitive benefits from its decision to require Microsoft to supply interoperability information to facilitate competition for workgroup servers and media players. The EC rejected Microsoft's defense that it had no obligation to license its protocols, some of which were protected by intellectual property rights, and stated:[55]

If competitors had access to the refused information, they would be able to provide new and enhanced products to the consumer. In particular, market evidence shows that consumers value product characteristics such as security and reliability, although those characteristics are relegated to a secondary position due to Microsoft's interoperability advantage. Microsoft's refusal thereby indirectly harms consumers.

The EC rejected the concern that obligations imposed on Microsoft to assist rivals by providing interoperability information would undermine innovation incentives and conflict with the objectives of intellectual property protection.

The US and Europe are an ocean apart regarding obligations of dominant firms to assist their rivals. US law is arguably too accommodating to conduct by dominant firms that excludes rivals, while European antitrust law fails to identify the conditions under which an obligation to assist rivals is procompetitive. Nonetheless, even if failure to support interoperability is not an antitrust *violation*, a requirement to support interoperability is a valid *remedy* for anticompetitive conduct.

The purpose of a remedy is to restore competition and deter future anticompetitive conduct. It is difficult to restore competition that might have occurred if Microsoft had not squashed the threat from Netscape and platform-independent middleware, both because the middleware threat was an exceptional opportunity to erode Microsoft's monopoly and because in an industry such as PC operating systems with powerful network effects, merely bringing an end to exclusionary practices is not sufficient to reverse ill-gotten gains. An interoperability requirement is a forward-looking remedy that prevents Microsoft from extending its unlawfully maintained dominance in client operating systems into workgroup servers. In theory, at least, it is one of the most important elements of the settlements that resolved the US and European antitrust cases. In practice, the EC and the plaintiffs in the US case encountered numerous difficulties regarding enforcement of these interoperability requirements.

Two years after the EC required that Microsoft supply complete and accurate interface information that would allow non-Microsoft workgroup servers to achieve full interoperability with Windows PCs and servers, the EC concluded that Microsoft had not provided sufficient technical documentation to allow competitors to develop interoperable servers. It also held that Microsoft's license fees were excessive, before the company reduced them in October 2007, three years after the EC settlement decree. The EC fined Microsoft 899 million euros for its noncompliance.[56]

A report on the effectiveness of the Final Judgment that settled the US case also alleged that Microsoft did not live up to its obligations to supply interoperability information.[57] The Final Judgment originally had a five-year term expiring in November 2007, but at the end of the decree's original term, a technical committee established to monitor compliance identified hundreds of instances in which Microsoft had not disclosed adequate documentation to guide interoperability.[58] In response to these and other concerns, Microsoft agreed to extend selected provisions of the Final Judgment governing the licensing of client-server communications protocols until November 2009. The presiding judge added another extension until May 2011.

The disclosure and licensing obligations included in the settlements of the US and European Microsoft litigation had the potential to restore competition and promote innovation for PC and workgroup operating systems. Unfortunately, they were too limited in scope to promote substantial competition and the requirements were not clearly specified, which led to lengthy and complex disputes over the required disclosures, the necessary degree of documentation, and reasonable licensing terms. It is possible that broader disclosure obligations would have been more effective and easier to enforce, with results more similar to the positive outcomes from compulsory licensing obligations in the 1956 AT&T and IBM consent decrees and the 1975 Xerox consent decree (see chapter 6).

### The middleware threat to Microsoft

A criticism of *US v. Microsoft* is that the case was built on a false premise: that Navigator and other middleware products would eliminate the applications barrier to entry that protected Microsoft from competition for operating system software. This presumption was present, albeit to a lesser extent, in the European litigation as well. More than a dozen years after the DOJ, in concert with the plaintiff states, and the EC settled their allegations against Microsoft, this promise has yet to be fully realized.

Furthermore, developments occurred in the PC industry that promote interoperability and do not require Java. Microsoft has continued to support the Office productivity suite for Apple computers as well as Windows-based PCs, and has supported protocols such as Open Office XML that allow computer users to open and save documents between Microsoft Office applications within the same platform and between platforms (such as Microsoft Word for Windows and for Apple operating systems). Open-source programs such as the Apache Open Office productivity suite also facilitate porting work product between different computer platforms.

Browsers have evolved and expanded functionality without providing standardized support for Java.[59] Google offers a suite of productivity applications that users can access with the Chrome browser from any platform that runs it.[60] Many other important applications are cloud-based, including products and services offered by Salesforce.com, Dropbox, and Adobe. There are also document creation tools, financial software, and websites for hosting software development tools and collaboration. These applications reside on remote servers and can be accessed by a client with little computing power. In this respect, they come close to the world without an applications barrier to entry envisaged by the plaintiffs in the Microsoft case.

Although the threat that Netscape and Java would eliminate Microsoft's grip on PC operating systems did not materialize, it was correct as an economic matter to hold Microsoft liable for its conduct. The company's restrictive contracts had little (if any) efficiency justification. Importantly, if Microsoft had been absolved of any antitrust liability, it could have continued its restrictive practices, which would have erected entry barriers to new competition from browsers and internet applications.

In summary, the antitrust cases filed against Microsoft in the US and Europe promoted competition and innovation in the computer industry even if they were based on flawed predictions about the evolution of the industry.

### Microsoft's efforts to fragment Java

The district court found that Microsoft sought to defeat industry adoption of a platform-independent Java by effectively splitting the Java standard. The court of appeals agreed that Microsoft harmed competition by deceiving Java developers about the cross-platform portability of its proprietary Java implementation and by coercing Intel to stop assisting Sun Microsystems to improve the Java technologies. It also affirmed that these actions supported a finding that Microsoft engaged in unlawful monopolization in violation of the Sherman Act. However, the court refused to affirm the district court ruling that Microsoft committed an anticompetitive act by designing a Java virtual machine (JVM) that was proprietary to the Windows platform. The court of appeals followed precedent in US antitrust law that even monopolies have broad discretion to choose their product designs and innovations. Moreover, the court of appeals found that Microsoft had an efficiency justification for its Java design because its efforts improved the performance of Java for Windows environments.

Based on the logic of the case and the objective to restore competition lost by Microsoft's conduct, the Final Judgment reasonably could have required Microsoft to offer support for Sun's Java implementation without prohibiting the company from promoting its own version of Java. The monopolization case against Microsoft focused on the company's efforts to prevent the Netscape browser and other Java-based middleware from breaking the applications barrier to entry. The promise of a nonproprietary Java platform was at the crux of the case. Had Microsoft not splintered the Java standard, it is possible that Java would have fulfilled its promise to become a platform for the development of easily portable PC applications.

### Rule of reason treatment for tying

Economics offers several arguments for why product tying and its close cousins, bundling and technological tying, can have anticompetitive effects. By requiring a customer to purchase both the operating system and the IE browser as a package, the benefit to a customer from a different browser is no greater than the incremental value of the alternative browser relative to IE.[61] This is how the tie cuts off a rival's "air supply" and makes rival entry more difficult if the rival has to charge a price to cover its cost. The net benefit from an alternative to IE could be even less than its incremental value if the customer has transaction costs to purchase and install the rival browser, or if disk space limitations or compatibility issues impose additional costs.

It is often more profitable for a firm to offer customers the option to purchase products separately rather than as a tied sale because this allows a firm to more profitably sort customers according to their willingness to pay for the products.[62] If a firm with monopoly power refuses to offer this option, it raises a plausible concern that the intent of the tie is to exclude rivals by making entry more difficult. Of course, this is hardly definitive evidence in support of monopolistic intent because it is common for firms that have no significant market power to offer products only as a bundle, and such firms could not success-fully employ bundling to monopolize a market.[63]

A related intent argument is that Microsoft should have had an incentive to work with, rather than against, Netscape because the Navigator browser was a product which, like other complementary applications, added value to the Windows operating system. The fact that Microsoft took numerous measures to impede Netscape suggests that Microsoft was more concerned that Navigator would create a path for competition with Windows rather than adding value to Windows.[64]

Similar arguments for and against pure bundling apply to technologically integrating the operating system and the browser. There are innocent reasons for a technological tie. Microsoft could have anticipated that nearly all consumers would want the browser with the operating system and may have viewed a bundled offering as a positive step that would satisfy consumer demand. Offering the browser separately would impose transaction costs and make it more difficult for consumers to access the internet to install products and upgrades. Furthermore, a separate browser product would incur additional product support costs. Today, the notion that operating systems should be available without internet browsers is as antiquated as the notion that word processing programs should not include spell-checkers or that cell phones should not include cameras.

These benefits from integration do not excuse Microsoft for engaging in practices that excluded competition from Netscape and other middleware, but they support the conclusion of the court of appeals that software bundling and product integration should not be treated as per se unlawful. The court correctly remanded the tying allegations for analysis under a rule of reason standard, although the Settlement preempted that evaluation.

## 5  Was the Final Judgment a Lost Opportunity?

The Final Judgment that resolved *US v. Microsoft* did not accommodate the plaintiffs' initial request to cleave Microsoft into separate operating system and application companies. Some believe that the divestiture would have created incentives for the independent applications company to port applications to competing operating systems and for the operating system company to facilitate competition in applications and middleware.[65] As an integrated supplier of operating systems, application software, and middleware, Microsoft has an incentive to exclude or disadvantage rival products.[66] The divestiture would have eliminated or significantly reduced the incentive for Microsoft to preference its own applications and middleware.

Although the more modest conditions in the Settlement elicited harsh criticism,[67] the proposed divestiture could have raised prices and diminished innovation and would have had large administrative costs. An independent supplier of the Microsoft Office productivity suite would not account for the positive effect of a lower price for the Office suite on the demand for the Windows operating system, and vice versa for the operating system company. A single company that sells both the operating system and the Office suite would take these positive interactions into account and likely choose a lower profit-maximizing total price for both products. This is the "Cournot complements effect" discussed in chapter 2. A similar Cournot complements effect would apply to innovation incentives for the operating system and complementary applications under reasonable assumptions about demand.[68]

If the Settlement had required the divestiture, courts would have had to interpret and enforce the restrictions on line-of-business activities, including the technical requirements that define an operating systems company, an application, and middleware. Howard Shelanski and Gregory Sidak maintain that interpretation and enforcement of the proposed structural remedy would have been at least as onerous as the interpretation and enforcement of the 1982 consent decree that provided the framework for the dissolution of the Bell System.[69] Furthermore, the telecommunications industry eventually recreated the vertically integrated structure of the pre-divestiture Bell System following the Telecommunications Act of 1996. It is likely that similar reintegration would have followed structural separation of Microsoft, with network effects driving businesses and consumers to support a dominant supplier.

Although a vertically integrated firm has an incentive to disadvantage its rivals, an independent monopoly supplier of the operating system can have an incentive to squeeze independent suppliers of complementary applications by charging a high price for the operating system or by investing aggressively in R&D to develop complementary products, if it is permitted to do so. An independent operating system monopolist has an incentive to invest aggressively in complementary applications because, even if its R&D fails to produce the best applications, it can create products that discipline market prices, which allows the firm to extract a higher price for the operating system. It is a case of "heads I win, tails you lose." Total industry innovation could suffer from an independent monopoly supplier's aggressive conduct.[70]

Integration of the operating system and applications facilitates coordination of R&D effort and promotes knowledge flows between complementary activities. Integration also can mitigate "holdup" that can occur when an independent firm makes R&D investments and subsequently bargains over the division of profits between the inventor and a user or other beneficiary of the invention. Holdup refers to strategic bargaining after the counterparty has made large and unrecoverable expenditures, such as investments in R&D. Because the investments are costs that have been spent in the past and cannot be recovered, the party that made these investments does not have a credible threat to reject a bargain merely because it does not compensate these costs. The risk of holdup in future negotiations can discourage independent R&D investment. An integrated firm does not have an equivalent risk of strategic bargaining.[71]

The Final Judgment could have been tougher on Microsoft, but the Settlement and the threat of continued oversight by US and European antitrust enforcers had beneficial consequences for competition and innovation. The prohibitions on restrictive agreements in the Final Judgment likely facilitated competition for internet browsers. Google Chrome is the most popular internet browser by a large margin[72] and delivers some of the promise of *US v. Microsoft* by enabling web-centric applications, many of which are powered by servers that run Unix and Linux operating systems. David Heiner, writing as Deputy General Counsel at Microsoft, explained a Microsoft commitment to interoperability embodied in a 2008 policy statement in part by the desire to attract developers to Microsoft's platform and in part by ongoing competition law scrutiny.[73]

Structural divestiture of Microsoft into separate operating and application companies might have eventually led to greater competition and more innovation in the PC industry, although that outcome is highly uncertain and divestiture would have created inefficiencies, at least in the near term, and would have had large administration costs. It is unfortunate that the district court approved the plaintiffs' proposal to break up Microsoft without a hearing. Consequently, there is no record to evaluate the merits and costs of the proposed divestiture, which led the court of appeals to vacate Judge Jackson's order and remand the case to the district court with instructions to consider a remedies-specific evidentiary hearing (which the Settlement preempted).

## 6 Lessons from US v. Microsoft

US v. Microsoft illustrates several themes that are generally relevant to competition policy for the high-technology economy and that appear repeatedly in this book in the context of other antitrust enforcement actions that allege harm to innovation or future price competition. I list several of these themes in this section.

### The importance of protecting nascent competition

The district court took a hard line about attempts to eliminate nascent competition from Netscape Navigator and Java, without regard to whether that competition was likely to be realized. Judge Jackson opined that "it is not clear whether, absent Microsoft's interference, Sun's Java efforts would by now have facilitated porting between Windows and other platforms enough to weaken the applications barrier to entry. What is clear, however, is that Microsoft has succeeded in greatly impeding Java's progress to that end with a series of actions whose sole purpose and effect were to do precisely that."[74]

The court of appeals agreed. The higher court explained:[75]

The question in this case is not whether Java or Navigator would actually have developed into viable platform substitutes, but (1) whether as a general matter the exclusion of nascent threats is the type of conduct that is reasonably capable of contributing significantly to a defendant's continued monopoly power and (2) whether Java and Navigator reasonably constituted nascent threats at the time Microsoft engaged in the anticompetitive conduct at issue. *As to the first, suffice it to say that it would be inimical to the purpose of the Sherman Act to allow monopolists free reign to squash nascent, albeit unproven, competitors at will—particularly in industries marked by rapid technological advance and frequent paradigm shifts.*

The court's condemnation of Microsoft's conduct without requiring evidence of actual or likely competition from Java and Netscape is an important precedent for antitrust enforcement in high-tech industries. Dominant firms are often able to identify nascent competitive threats in these industries and eliminate them before they mature into significant competitors. The appellate court in US v. Microsoft confirmed that the antitrust laws can block conduct that threatens competition in dynamic industries without requiring plaintiffs to establish with a high degree of certainty that the threatened competition would have had a significant effect on market outcomes. That precedent can reasonably apply to acquisitions by dominant firms of potential competitors, despite the absence of a track record to establish that the acquired firms would be significant rivals if they remained independent.

### Innovation is hard to predict

This may be obvious, but antitrust enforcers should not forget that innovation evolves in unexpected ways and from unexpected sources. Innovations change the ways that firms compete and frustrate attempts by courts to define markets that describe where future competition is likely to occur.

The computer industry evolved in ways that the DOJ and plaintiff states did not envisage in their complaint. Apple is a more significant competitive force than it was in 1998, and PCs face new competition from smartphones and other portable devices. Consumers no longer purchase browsers at computer stores. Cloud computing has transformed the industry by providing a remote server-based platform for applications that is agnostic to the client's desktop operating system. Cloud computing has achieved some of the objectives of the Microsoft antitrust litigation by moving applications off the desktop. But cloud computing has not commoditized the operating system, and it owes its success more to industrywide internet protocols than to adoption of a common Java technology. Events did not prove that the applications barrier to entry for PC operating systems would have evaporated if Microsoft had not "cut off Netscape's air supply."

Nevertheless, and despite the fact that the antitrust investigations of Microsoft may have lagged the pace of the computer industry, there were valid reasons to challenge Microsoft's conduct. A virtue of the Microsoft antitrust litigation is that the case proved to be robust to different futures for the PC industry. Browsers have not replaced operating systems, but they have evolved in their own right to become a valuable and diverse product category. Microsoft's conduct, if left unchecked, could have harmed consumers in many ways, such as by constraining the ability of software developers to support other computer platforms and by enforcing proprietary protocols to defeat interoperability. It is unlikely that firms and consumers would have the spectrum of choices they enjoy today if the courts had not constrained Microsoft's behavior.

### Conventional analysis is suitable for some monopolization cases that involve innovation

Both the district court and the court of appeals identified innovation as a central concern in the Microsoft case, but the case mostly turned on classic antitrust jurisprudence for enforcement of the Sherman Act. The courts did not explicitly consider whether innovation concerns justify greater tolerance for restrictive practices by dominant firms or whether they amplify conventional concerns that focus on price effects, with the exception of special treatment for the tying of software products.

Fortunately, innovation concerns do not typically justify exceptional treatment for monopolizing conduct. In industries such as computer software, dominant firms benefit from entry barriers that insulate them from price and innovation competition. Examples of these barriers include brand-specific network effects, such as the Windows ecosystem, which rewards loyalty to Microsoft's software products. Patent protection, economies of scale, and reputation also work to protect established firms from competition. Artificial entry barriers, such as exclusive dealing arrangements and intentional system incompatibilities, further shield established firms in these industries from price and innovation competition.[76] While there may be circumstances in which artificial entry barriers have efficiency benefits, a reasonable starting presumption for competition policy is that such barriers are not necessary to promote innovation. There are many other ways for firms to capture value that do not require conduct that excludes rivals.

The Arrow replacement effect teaches that an incumbent firm's profits deter investment in a new product that would replace the firm's existing profits. Weak antitrust enforcement is at least as likely to increase an incumbent's profits from its existing products as it is to increase profits from new products. Therefore, it is likely that the net effect of weaker antitrust enforcement would be to increase the Arrow replacement effect and deter innovation by an established firm.

Firms that are new to an industry have stronger innovation incentives than established firms if they can obtain comparable benefits from successful inventions. New entrants do not have profits that are at risk from innovation: They do not suffer from the Arrow replacement effect. Established firms can neutralize new entrants' strong innovation incentives if they engage in strategic conduct, such as exclusive dealing or predatory pricing, that denies entrants the ability to profit from their inventions.

These arguments bolster the case for strong antitrust enforcement to prevent exclusionary conduct by dominant firms that would threaten innovation. Of course, there are exceptions, but a good first approximation is that exclusionary conduct by a monopolist that would harm consumers in the absence of innovation concerns is also likely to harm consumers by reducing the incentives of the monopolist to innovate.[77]

**Platform competition does not always necessitate a two-sided market analysis**

The Windows operating system is a platform that connects computer users to computer applications and device manufacturers. Nonetheless, the Microsoft litigation did not turn on two-sided economic issues. Microsoft's exclusionary contracts, its tying and integration of the operating system and the browser, and its splintering of the Java standard were about preventing a potential competitive threat from middleware. It is unlikely that a careful consideration of two-sided interactions would have uncovered platform interaction effects that would justify reversing the courts' conclusions. Two-sided platforms raise important issues related to complementary activities, but not every litigation involving a platform market requires a two-sided analysis.

**There is a fine line between complementors and competitors**

Complementary products and services add value to a firm that sells or licenses a complementary product or service, but complements can also create a path for new competition. That was Microsoft's worry about Navigator. The Microsoft case exposes one of many flaws in the theory of "one monopoly profit" advanced by conservative antitrust enforcers. Under that theory, the monopoly supplier of a product, such as an operating system, has no incentive to exclude competition from efficient suppliers of complementary products (such as software applications or workgroup servers), because the monopolist can charge a price that extracts the contribution made by the complementary products to the value of the combined system.

The one monopoly profit theory fails when the complementors are a potential threat to the monopolist's market dominance. Netscape threatened Microsoft's dominance with a browser product that could break the applications barrier to entry. Workgroup servers are a threat because they can marginalize the desktop operating system by running applications on a network. The one monopoly profit theory can fail for other reasons. For example, competition can interfere with the monopolist's ability to charge usage-dependent prices, and a monopoly supplier can have inefficient incentives to invest in R&D for complementary products.[78]

Concerns about competition from complementors have appeared in other antitrust cases. Examples include a number of older cases in which firms sold peripheral equipment, such as disk drives and printers, that were plug-compatible with IBM mainframes, meaning that they could be easily connected to and interoperate with the mainframe. These components added value for IBM, but they also provided entry points for competitors and limited IBM's discretion to impose usage-dependent prices.

### Standards and interoperability are critical for innovation but difficult to regulate

The Microsoft cases raise two separate issues related to standards and interoperability. One is the allegation that Microsoft engaged in conduct that splintered the Java programming language. The appellate court held that Microsoft's conduct with respect to Java was anticompetitive, but the court did not require Microsoft to support a common implementation of the Java technologies. The court's opinion reflected a reluctance to regulate adherence to industry standards and is consistent with a view that competition policy should not prevent firms from developing their own solutions, even if those solutions cause a standard to fragment. That policy hinges more on the difficulty of regulating compliance with an industry standard than on the desirability of an industrywide standard. The court could have obligated Microsoft to support a common Java standard without preventing Microsoft from developing its own Java implementation. The EC followed this type of approach when it compelled Microsoft to offer a ballot box that allowed consumers to choose among different browsers.

A second standards-related issue is the requirement in the Final Judgment that settled the US case and in the decision by the EC that Microsoft support interoperability between Windows operating systems and non-Windows server operating systems and other products. Although the US case did not challenge conduct by Microsoft that impeded server interoperability, the obligation to disclose server-related information and to license related intellectual property at reasonable and nondiscriminatory terms is an important and forward-looking element of the Final Judgment.

The compulsory licensing obligations in the Final Judgment and the EC decision did not warrant concern that they would severely undermine Microsoft's innovation incentives. The protocols and related information at issue do not allow a firm to clone a Microsoft product. Rather, they provide information necessary to facilitate a large ecosystem of complementary products and services. Unfortunately, these obligations turned out to be the most difficult conditions to enforce. One difficulty was the lack of experience in documenting and licensing technology necessary for interoperability. While Microsoft had a history of making APIs available to programmers (because it was in its interest to do so), there was no corresponding history for documenting and licensing interoperability technology. Problems arose in identifying as well as adequately documenting the required technologies, defining the breadth of licenses, and determining reasonable royalty structures.[79]

Experience with the interoperability requirements in the Microsoft cases has implications for enforcement actions in other cases. Enforcement of obligations to supply interoperability information and relevant intellectual property is administratively difficult unless the obligations are specified very clearly, or they are so general that they do not raise questions about the information and intellectual property licenses that must be supplied or the scope of the requirements.

### The ease of integrating into related technological areas makes these cases difficult

The ease with which firms that hold a dominant position in a digital technology can extend their dominance by integrating into related markets is a vexing challenge for antitrust enforcement. In many industries, integration into a related market involves large investments and continuing costs to manufacture, market, and distribute the related products. Computer software and digital services more generally are different. While software vendors can have large upfront costs to develop a new product, the incremental costs of producing, marketing, and distributing the product can be very low. It is not costly to bundle a media player, paint program, or spell-checker with the release of a software product such as Windows or Office. It was also a relatively low cost for Microsoft to include IE on the same CD with Windows 95 and to integrate code for IE into Windows 98 and subsequent operating systems. Consumers can benefit from the low cost of distributing the new product, but the integration can also foreclose actual and potential rivals in the new market, diminish competition, and harm subsequent innovation.

Many tech platforms can scale up easily and expand rapidly into new activities at relatively low incremental cost. Because many tech platforms can easily integrate into new applications, they can often copy start-ups that attempt to compete, which deters independent innovation unless the intent is to sell the innovation to an incumbent platform. The ease of integrating into related technological areas complicates antitrust enforcement that seeks to prevent dominant firms from monopolizing related markets. It is often difficult for antitrust enforcers to determine when related activities constitute separate markets in which competition may be harmed and difficult for enforcers to weigh the costs of monopolization against the benefits of having an established firm supply new products. If enforcers determine that a dominant firm has monopolized a related market, it can be difficult for enforcers to fashion a workable remedy.

The next chapter examines the tradeoff between innovation by a dominant firm and harm to competition in the context of the design of Google's internet search engine, and highlights the difficulty of implementing effective remedies for alleged anticompetitive conduct related to product integration and design. The Microsoft case may be old news, but it reveals issues that continue to challenge antitrust enforcement for the high-technology economy.

# 9 "Where Is Foundem?": The Google Shopping Case and Antitrust Policy for Product Designs

> Product innovation generally benefits consumers and inflicts harm on competitors, so courts look for evidence of exclusionary or anticompetitive effects in order to distinguish between conduct that defeats a competitor because of efficiency and consumer satisfaction and conduct that impedes competition through means other than competition on the merits.
> —*New York v. Actavis PLC*, US Court of Appeals (2015)

## 1 Introduction

This chapter discusses competition policy for product designs and related new product offerings by a dominant firm. I introduce this topic with a discussion of investigations by the US Federal Trade Commission (FTC) and the European Commission (EC) into allegations that Google manipulated search results to benefit some of its proprietary products.[1] The central issue in those investigations was the design of Google search algorithms that allegedly discriminated against independent comparison shopping services (CSS). Comparison shopping websites collect information from participating merchants and the World Wide Web about product prices, features, and reviews, which they monetize with advertising or marketing agreements. They include Foundem, a UK comparison shopping website that gained notoriety as the lead plaintiff in an antitrust suit brought by the EC, and other websites such as NexTag, PriceGrabber, Shopping.com, and Shopzilla.

Google launched a CSS called Froogle at the end of 2002, which it rebranded as Google Product Search and then Google Shopping in 2012. Over the course of these developments, Google made changes to its search algorithms, which order results on its search engine results page (SERP). Prior to the changes, a product query would deliver links to non-Google CSS websites at or near the top of the SERP. After the algorithm changes, a product query would display Google product listing ads at or near the top of the SERP, complete with images and price data, and links to third-party CSS websites would appear on distant SERPs, if they appeared at all. Hence the title of this chapter: "Where Is Foundem?"

The Google Shopping case proved to be a Rorschach test for antitrust enforcement. The FTC saw a dynamic company that made legitimate product improvements and thus terminated its investigation, with no conditions related to Google's alleged display bias. The EC, however, saw a dominant firm that excluded rivals. The EC fined Alphabet (Google's parent) and ordered the company to end its alleged discriminatory practices.

The Google Shopping case is important because it addresses competition policy for product designs in the high-technology economy and does so in the context of a platform market. Section 2 briefly describes the two-sided market for internet search. Section 3 addresses the evaluation of market power for the so-called free side of this market. Consumers do not pay a financial price to query the internet, but their search histories supply Google and other search engines with valuable data that they can use to attract paying advertisers.

Section 4 reviews the disparate outcomes of the FTC and EC investigations in this case. The FTC did not provide a detailed explanation of its decision to absolve Google from antitrust liability for alleged bias in the display of its search results. The EC issued a lengthy decision but did not tether it to general principles that could guide competition policy for product designs or related innovations.

Section 5 discusses general approaches to evaluate product designs and other innovations by a dominant firm that may exclude competition. This section reviews several proposals that have been advanced by courts and antitrust scholars and concludes by supporting a truncated rule of reason approach. The truncation refers to a threshold level of innovation above which a rule of reason balancing is not necessary. Under this truncated rule of reason, product designs and other innovations would escape antitrust condemnation if they were substantial improvements and were not accompanied by other exclusionary conduct that did not have procompetitive benefits. Innovations and product designs that provide only marginal benefits would be examined under the rule of reason to determine whether their benefits compensate for any exclusionary effects. This approach would condemn some product changes that have occurred in the pharmaceutical industry, where the regulatory environment allows the suppliers of patented drugs to exclude rivals by making minor changes to their products. Section 6 offers some concluding remarks for competition policy for innovation and product designs.

## 2   A Brief Primer on Internet Search

Search engines enable consumers to explore the billions of pages of content on the World Wide Web and allow merchants to place ads to attract potential customers. Without a search engine, the web would be like a giant library with no card catalog. An efficient search engine is indispensable to access and utilize the information on the internet. A search engine is also a valuable tool for advertisers to connect with consumers and exploit opportunities for e-commerce.

Google operates a two-sided platform that allows internet users to query billions of web pages without a financial charge (they compensate Google by surrendering valuable personal data to Google), while providing paid services that allow advertisers to place their ads on the SERP or on the pages of website publishers. A Google search query delivers "organic" results (also called "generic" results), which are links to websites that the Google search algorithms conclude are relevant to the query. In addition to organic search results, a query may trigger advertisements that are displayed alongside or nearby the organic results on the SERP. Advertisers compete for favorable placement on the SERP by placing price-per-click bids on query keywords. Google ranks the bids based on the bid prices and the likely number and quality of clicks on the ads, which generate revenue for Google.[2]

For several years, most search engines displayed a combination of organic search results and text ads in response to search queries. But this "ten blue links" model began to change when Google and other search engines evolved from delivering relevant website links to delivering relevant information. Instead of supplying only website links in organic search results in response to a query, Google and other search engines evolved to, in some cases, answering the question itself, along with relevant images and text.[3] Google introduced a functionality called Universal Search in 2007, which integrated specialized displays in search results that can include product shopping listings and other results, such as flight, restaurant, and hotel listings, maps, and news.

The specialized displays, such as Google Maps, are sometimes called "verticals" because they focus on a segment of online content. Verticals may be proprietary products (such as Google Flights), and they may or may not include sponsored (paid) links. For example, a search for "Italian restaurants Boston" may deliver a map that includes restaurant locations, some of which may be sponsored.

Many website publishers and online merchants depend on favorable rankings in Google's free organic search display for a significant fraction of their traffic. Changes to the Google search algorithms that push more of Google's specialized content to the top of the SERP necessarily denies to others premium SERP real estate for organic search results. Most internet searchers focus on the first or second pages of search results, and one estimate is that exclusion from the first five pages of search results leads to a 90 percent reduction in organic clicks.[4] Of course, consumers focus on websites that rank high in search results in part because they believe, often correctly, that they are most responsive to their queries.

The Google search engine is not a static product. Google makes numerous changes to its search algorithms. Relatively minor tweaks occur frequently—one estimate is that 500–600 changes per year.[5] Sometimes Google makes major changes. A few years after it introduced its Universal Search display, it undertook an effort to identify and demote search results that it believed offered little or low-quality content, including sites that primarily outsource content to third parties or have high ad-to-content ratios. Google formalized this change in the "Panda" update, which went live in early 2011.[6]

The Panda update greatly reduced traffic for CSS websites such as Foundem and related websites that mostly aggregate content from other sources. The update identified these websites as being of low quality because they offered little original content; they were mainly a collection of links to other websites, such as retailers for cameras. After the Panda update, relevant queries no longer returned links to CSS and similar aggregation websites on the first or second SERP. They were more likely to appear on much more distant pages, if at all.

The purported intent of the Panda update was to give greater prominence in organic search results to high-quality websites, which by itself, would seem to benefit internet users.[7] However, the Panda update did not demote Google's product listing ads despite the fact that they did not offer much original content and in most respects were similar (and arguably inferior) to other CSS. To the contrary, after the Panda update,

Google displayed ads with links to Google Shopping prominently on the first SERP in response to a product-related query.

If Google had relied solely on organic search results to display Google Shopping, the Panda update would have buried Google's own CSS in the hinterland of organic search engine results, along with other aggregation websites that offered little original content. A Google employee wrote:[8]

From a principal perspective it would be good if we [Google] could actually just crawl our product pages and then have the[m] rank organically (...) Problem is that today if we crawl [our product page] will never rank.

Google's preferential treatment for its own comparison shopping vertical prompted many CSS publishers and some other suppliers of specialized web content (such as Yelp and TripAdvisor) to cry foul and complain to the FTC and the EC. Was Google's conduct anticompetitive? And even if it did harm competition, was this an unavoidable consequence of a design change for which Google had a valid business justification?

An investigation into anticompetitive conduct typically begins with an inquiry into whether the defendant has significant market power. Market power is a screen for possible anticompetitive effects; if a firm does not have market power, it does not have the ability to profitably raise prices or exclude competition. Market definition also plays a role in assessing whether Google has anticompetitively extended the market power that it may possess in general internet search into a separate market for CSS. In the two-sided market for internet search, anticompetitive effects can include higher prices paid by advertisers; a reduction in the quality of internet search; or a reduction in innovation or product quality. The next section briefly addresses whether Google has sufficient market power in general search to exclude competition in CSS.

## 3 Does Google Have Market Power in Organic Search?

Google does not charge consumers to search the web (although the company profits from collecting information about consumers' search behavior) and its conduct did not raise the consumer price for search above zero. It is important at the outset to dispel a popular myth that a zero price implies zero consumer harm. Google charges a zero price for search because doing so allows Google to make more money from advertising. Google's conduct can impose a quality cost on consumers if it makes search results less informative. Alternatively, consumers can be harmed indirectly if the conduct increases the cost of advertising, which consumers ultimately bear. Consideration of the latter indirect harm involves a two-sided analysis.

Google is the most popular search engine by a very large margin. One estimate is that in 2018, searches on Google accounted for more than 87 percent of page views from all platforms (i.e., desktops, mobile devices, and tablets) in the US, and more than 93 percent of page views from all platforms in Europe.[9] Nonetheless, Google maintains that it cannot have market power necessary to lower search quality or exclude competition because "competition is only a click away."[10] If a consumer is not satisfied with a Google search result, the consumer can easily turn to a different search engine, such as Yahoo!, Bing, or DuckDuckGo, all of which have capacity to respond to additional queries, or the consumer can navigate directly to a relevant website. A further claim that Google lacks market power in search is that the advertising side of the two-sided market for internet search encourages Google to supply high-quality search results. Advertisers pay Google because its search engine identifies relevant potential customers. Advertisers would pay less if Google's search results were less informative.[11]

These arguments have theoretical appeal, but they do not answer whether Google actually has the incentive and ability to degrade the quality of search results. The company has the ability to degrade internet search results without causing an unprofitable loss of consumers for several reasons. First, internet search has aspects of a credence good, which by definition is a good whose quality is difficult for a consumer to verify.[12] For complex queries, it is difficult for a consumer to know whether Google has returned an accurate result or whether a different search engine would have returned a better result.

Second, Google's enormous trove of search results and its technological expertise allow it to provide more accurate responses to queries compared to other search engines, even if those responses may be somewhat distorted by financial incentives. The search engine DuckDuckGo touts that it does not track consumers, which arguably implies that it is more interested in consumer privacy than in providing search results that generate advertising revenue. But DuckDuckGo processes only a tiny fraction of search queries compared to Google, which limits the ability of its search algorithms to supply accurate and highly relevant responses when consumers enter queries that require a complex evaluation.[13]

A third reason why Google has the ability to profitably degrade search results for comparison shopping is that a product query is only one of many categories of queries that consumers enter into a search engine. Google could degrade the quality of search results for products without causing many consumers to become generally dissatisfied with the Google search engine if they place a low value on product queries relative to the total value that they obtain from all other Google search queries.

These reasons support a conclusion that Google has the ability to degrade search results without causing a significant exodus of consumers to rival search engines and that Google's market share reflects significant market power for organic internet search.[14]

Market power gives Google the ability to profitably degrade search results, but does Google have the incentive to do so? Economic theory suggests that firms that have a monopoly on an important input (such as internet search) and compete with other firms in markets that employ the input (such as CSS websites) have incentives to degrade the quality of the input that they supply to their rivals.[15] Organic search results that provide users with the best information do not necessarily maximize a search engine's profit if the search engine has proprietary services (such as a shopping vertical) that compete with desired organic search results or if the search engine can promote organic search results that are likely to direct consumers to websites where the search engine can collect substantial advertising revenues.[16] The incentive for Google to accomplish an anticompetitive end by degrading the quality of its search results can be high because Google does not profit directly from search queries; instead, Google profits from advertisers that respond to search queries and from advertisements on its proprietary services.

A related question is whether there is a separate antitrust market for CSS. The EC concluded that CSS is a separate market for antitrust analysis, while the FTC did not address this issue in its commentary. Given the rapidly evolving nature of internet search, it is questionable whether it is useful to define a narrow market that is limited to a particular type of information, although the US Department of Justice defined a separate market for "comparative flight search services" in its investigation of Google's acquisition of ITA Software.[17] The assumption of a separate market for CSS makes it easier to conclude that Google's conduct harmed competition. If comparison shopping queries were like other queries for the purpose of antitrust analysis, the relevant question would be whether Google harmed competition for general search. That is an unlikely outcome because the publishers of CSS websites are not serious threats to Google's dominance of general internet search.

Google has the ability to display its own Google Shopping vertical service prominently and demote CSS websites without causing a significant exodus of consumers, and it likely has an incentive to demote a CSS because it competes with Google for advertising dollars. Does it follow that Google harmed competition without a procompetitive justification by designing an algorithm that demoted CSS in organic search results, while elevating its own shopping service to the top of the SERP? The EC answered in the affirmative, while the FTC reached the opposite conclusion. After briefly summarizing these outcomes in the next section, the subsequent section turns to general principles to evaluate product changes by dominant firms.

## 4   Two Antitrust Regimes, Two Different Outcomes

On January 3, 2013, the FTC voted unanimously to close its investigation of Google's search practices without demanding any change in conduct regarding the display of Google search results.[18] The Commission's closing statement emphasized that Google's design choices were an improvement:[19]

The totality of the evidence indicates that, in the main, Google adopted the design changes that the Commission investigated to improve the quality of its search results, and that any negative impact on actual or potential competitors was incidental to that purpose. While some of Google's rivals may have lost sales due to an improvement in Google's product, these types of adverse effects on particular competitors from vigorous rivalry are a common byproduct of "competition on the merits" and the competitive process that the law encourages.

Although the Commission acknowledged that "some of Google's algorithm and design changes resulted in the demotion of websites that could, collectively, be considered threats to Google's search business," it focused on the procompetitive effects from Google's design changes. The Commission stated:

Product design is an important dimension of competition and condemning legitimate product improvements risks harming consumers. Reasonable minds may differ as to the best way to design a search results page and the best way to allocate space among organic links, paid advertisements, and other features. And reasonable search algorithms may differ as to how best to rank any given website. Challenging Google's product design decisions in this case would require the Commission—or a court—to second-guess a firm's product design decisions where plausible procompetitive justifications have been offered, and where those justifications are supported by ample evidence.

The EC continued its investigation after the FTC decision. In June 2017, the Commission held that Google abused its dominant position in general internet search by giving illegal advantage to its own CSS, in violation of European Union (EU) antitrust law. The Commission fined Google's parent 2.42 billion euros ($2.7 billion at the prevailing exchange rate at the time) and ordered the company to cease discrimination against independent CSS.[20]

The Commission premised its decision on the following findings:

(i) General internet search services and comparison shopping services are separate relevant product markets for the purpose of antitrust analysis.[21]

(ii) Google has a dominant position in general search.[22]

(iii) Google abused its dominant position in general search by conduct that decreases traffic from Google's general search results pages to competing comparison shopping services and increases traffic from Google's general search results pages to Google's own comparison shopping service.[23]

(iv) This conduct is capable of having, or likely to have, anticompetitive effects in the markets for comparison shopping services and general search services.[24]

The EC also rejected justifications that Google advanced for its conduct. These included the consumer benefits from demoting low-quality websites, the value to consumers of providing the most useful and relevant search results, and the importance to Google of being able to monetize space on its SERP. The EC concluded that none of these purported justifications required Google to give preferential treatment to its own CSS.[25]

The EC's decision addressed the responsibility under EU antitrust law for a dominant firm to refrain from conduct that is an abuse of its dominance. Abuse of dominance under EU law encompasses a wide range of behavior other than competition on the merits that can hinder the degree of competition in a market or the growth of that competition.[26] Article 102 of the Treaty on the Functioning of the European Union specifically includes as an abuse of dominance "applying dissimilar conditions to equivalent transactions with other trading parties, thereby placing them at a competitive disadvantage."[27] It is an understatement to say that this is a weaker standard than US courts have adopted with regard to a monopolist's obligations to deal with rivals.

The EC decision, which exceeded 200 pages, concluded that Google's conduct has the potential to foreclose competing CSS, which may lead to higher fees for merchants and higher prices for consumers. It also concluded that Google's preferential treatment for its own shopping vertical reduces the incentive of competing CSS to improve the relevance of their existing services and create new types of services, and reduces the incentives of Google to improve the quality of its CSS because it does not need to compete on the merits.[28]

The EC decision emphasized discrimination and paid little attention to whether Google's algorithmic changes to its search displays were improvements. It stated that it "does not object to Google applying rich features to certain results but to the fact that Google applies such rich features only to its own comparison shopping service and not to competing comparison shopping services."[29]

The EC ordered Google to stop its discriminatory practices or else face continuing fines, but the Commission did not specify a remedy. Instead, it stated, "It is for Google and Alphabet, and not the Commission, to make a choice between the several possible lawful ways of positioning and displaying competing comparison shopping services in the same way as Google positions and displays its own comparison shopping service in Google's general search results pages, thereby bringing the infringement to an end."[30] This proved to be easier said than done. The EC rejected several proposals by Google, and the most recent proposal continued to raise Commission concerns as of March 2019.[31]

The Google Shopping case is not unique. Google's algorithms return specialized responses to many search queries. A query for "flights from Boston to Denver" shows the Google Flights specialized service with related booking options, such as hotels and car rentals, and may include sponsored links to other websites. Firms such as Yelp, TripAdvisor, Travelocity, Kayak, OpenTable, and Hotels.com can (and many do) complain that they are excluded from these services and may demand equal visibility for their websites. The EC opened an investigation into the Google Jobs search tool in response to complaints by rival job-search websites.[32] There are practical obstacles to accommodating these demands because there is a limited supply of premium real estate on a SERP.

The focus of this chapter is on the detection of anticompetitive product design rather than the design of appropriate remedies. It is not difficult to see why the FTC and the EC reached different outcomes based on their different legal regimes. The decision by the EC followed the road map described previously to assess abuse of dominance. The FTC decision is consistent with the deference to innovation in US antitrust law and legal precedents which support an enforcement posture that firms have little obligation to assist their rivals. Neither the FTC nor the EC described a general framework to evaluate the costs and benefits of design changes that have exclusionary consequences. The next section reviews several important antitrust cases in the US that dealt with exclusionary design changes by dominant firms, and discusses general principles that have emerged from these cases and from scholarly debate on this topic.

## 5   Antitrust Policy for Exclusionary Product Designs

US courts tend to defer to design changes that have technical merit if they are not accompanied by other exclusionary conduct.[33] In the words of the court of appeals in *US v. Microsoft*, "As a general rule, courts are properly very skeptical about claims that competition has been harmed by a dominant firm's product design changes."[34] Nonetheless, antitrust challenges to exclusionary product designs have survived, even under the accommodating principles of US antitrust law.

Several general approaches have been proposed to identify innovation and associated conduct that harm competition (sometimes called "predatory innovation"):

- A focus on whether a dominant firm coerces firms or consumers to adopt a new product (a hard switch)
- The profit-sacrifice test and its cousin, the no economic sense test (NEST), which ask whether the alleged conduct could be a rational business strategy if it did not harm competition
- A rule of reason analysis that compares anticompetitive effects and procompetitive justifications

All these approaches have both utility and limitations for evaluating innovation as a possible antitrust offense. I discuss each of them in turn.

### The hard switch

US courts addressed alleged anticompetitive product designs in the early 1970s in a case brought against the Eastman Kodak Company. The case involved, among other issues, the introduction of Kodak's Pocket Instamatic camera and a new color print film, Kodacolor II. The plaintiff, Berkey, competed with Kodak in supplying photofinishing services and cameras. Berkey alleged that Kodak violated the antitrust laws by failing to release advance information about the new film and camera format and by restricting Kodacolor II to the Instamatic format for a period of time, thereby preventing Berkey from providing photofinishing services or competing to sell cameras in the new format. The court of appeals reversed a jury verdict in favor of Berkey on these issues. The court ruled:[35]

If a monopolist's products gain acceptance in the market,... it is of no importance that a judge or jury may later regard them as inferior, *so long as that success was not based on any form of coercion.*

Other courts that have addressed allegations of anticompetitive innovation have distinguished between hard and soft switches. A "soft switch" occurs when a firm introduces a new product (such as a new camera format) but does not remove the old product from the market. In a "hard switch," the firm takes aggressive measures to remove the old product from the market or otherwise make it difficult for consumers to purchase the old product. For example, a drug manufacturer may reformulate a drug to make the dosage once a day instead of twice a day and then remove the old drug from the market so that it is no longer available for generic substitution.

Courts have used lack of coercion as a screen to identify innovation that does not raise antitrust concerns, much as they use small market shares as a screen to identify the absence of market power. The principle is that absent coercion, consumers will choose a new product or service only if it is better than alternatives. If there is conduct that facilitates a hard switch, it does not automatically follow that the conduct is anticompetitive. That should be assessed using a more thorough analysis of costs and benefits.

A focus on coercion frees courts from the difficult task of evaluating the merits of design changes; instead, it allows them to evaluate conduct (the hard switch) that may not have procompetitive justifications. However, coercion is a weak test to evaluate product changes because it is neither necessary nor sufficient for anticompetitive innovation. Firms can exclude competition without satisfying the literal test of a hard switch. Pharmaceutical companies spend heavily on research and development (R&D), but many companies spend more on sales, marketing, and administration (see table 9.1).[36] Much of this expense goes toward promotional activity directed to physicians and consumers. Drug manufacturers can switch patients to a new drug by promoting the new drug heavily while abandoning promotion of the old drug. That may not qualify as coercion, but it could have an equivalent effect.[37]

Minor changes to product characteristics can have exclusionary effects in the absence of coercion in some special circumstances. In other circumstances, a hard switch can have procompetitive benefits. When products have strong network effects, firms and consumers can

**Table 9.1**
Marketing and R&D expenditures by large drug companies (2016 $ billions).

| Company | Sales, Marketing, and Administrative Expenses | R&D |
|---|---|---|
| Johnson & Johnson | 19.0[a] | 9.1 |
| Novartis | 12.0[b] | 9.0 |
| Pfizer | 14.8 | 7.9 |
| GlaxoSmithKine | 14.1[c] | 5.4[c] |
| Merck | 9.8 | 10.1 |

[a] Excludes shipping and handling costs
[b] Excludes general and administrative expenses
[c] Converted from British pounds at 1 GBP = 1.5 $US
Source: Annual Reports.

exhibit excess inertia: They are reluctant to transition to a potentially superior but incompatible new product because they do not benefit unless other firms and consumers make the same adoption decision. On the supply side of the market, economies of scale can make it efficient for a firm to narrow consumer choices. Efforts to compel firms and consumers to adopt a potentially superior product can have procompetitive consequences when markets have these characteristics. Furthermore, some improvements have social values that greatly exceed their private returns.[38] A policy that ignores these spillover benefits and evaluates improvements based solely on their benefits for a narrow set of consumers could deter socially valuable innovation.

The existence of conduct that is unrelated to the product change and affects consumer and firm adoption decisions can aid fact finders in their assessments of harms and benefits. Coercive conduct that allows a dominant firm to maintain or extend monopoly power and has no procompetitive justification can be condemned as anticompetitive. In some cases, however, the presence or absence of coercion is not even an available dichotomy because design or other technological changes require rivals to adjust to the change to remain viable. In the digital economy, some firms supply automatic software downloads that can defeat interoperability.[39] Suppliers of complementary products must reconfigure their products if a firm that controls interoperability protocols implements a different technology. The change to the protocol itself coerces adoption of a new technology.

In the late 1960s, several companies sold devices such as disk drives and printers that were plug compatible with IBM's highly successful

360 family of mainframe computers. IBM might have welcomed plug-compatible manufacturers (PCMs) because they were complements and could add substantial value to a mainframe computer system. But independent pricing by PCMs interfered with IBM's usage-dependent pricing policies and peripherals were a potential threat to IBM by providing a stepping stone to mainframe system competition.

IBM fought PCMs with price reductions and implemented design changes to controllers and interconnections in its new 370 series of computer systems that had some performance benefits but defeated interoperability between the mainframe and non-IBM peripherals. Several PCMs responded with allegations of antitrust violations. In nearly all these cases, the courts held that where IBM's design changes had technical benefits, they were not anticompetitive.[40]

Four decades after the IBM cases, plaintiffs in the *in re Apple iPod iTunes Antitrust Litigation* made similar allegations regarding product designs that defeated interoperability. The plaintiffs alleged that Apple repeatedly updated its FairPlay encryption protocol to make its iPod media player and songs downloaded from its iTunes music store incompatible with other media players and streaming services, and also refused to license the updates to its rival, RealNetworks.

With regard to the refusal to license, the court cited the Supreme Court's 2004 decision in *Verizon v. Trinko*, with its proposition that a firm has no duty to assist its competitors.[41] As for the FairPlay updates themselves, the court focused on whether they were genuine improvements in response to hackers who had circumvented the FairPlay encryption. It answered in the affirmative for one of the updates and dismissed the charges. With regard to another update, the court could not determine whether it was a genuine product improvement and designated the issue for trial.[42] A jury concluded that the update was a genuine improvement and found no antitrust violation.[43] The court did not address whether coercion was relevant to the antitrust analysis. Apple implemented its FairPlay updates automatically, so consumers had no choice other than to accept the new encryption standards if they wanted to access the iTunes music store. According to the judge and jury, if the updates were improvements, they were sufficient to escape antitrust liability.

These cases imply that legitimate product improvements do not violate US antitrust laws if they are not accompanied by other exclusionary conduct.[44] In *Allied Orthopedic Appliances v. Tyco Health Care Group*, the court stated:[45]

There is no room in this analysis for balancing the benefits or worth of a product improvement against its anticompetitive effects. If a monopolist's design change is an improvement, it is "necessarily tolerated by the antitrust laws," unless the monopolist abuses or leverages its monopoly power in some other way when introducing the product. To hold otherwise "would be contrary to the very purpose of the antitrust laws, which is, after all, to foster and ensure competition on the merits." If a monopolist's design change is an improvement, it is "necessarily tolerated by the antitrust laws."

Although many courts have accepted the principle that that a design change cannot violate the antitrust laws if it is an actual improvement, as a standard to evaluate allegations of anticompetitive innovation it is too permissive. Suppose that IBM made changes to its interoperability protocols in a way that had a trivial technical benefit but excluded PCMs. Or suppose that Apple's FairPlay updates were intended to exclude rival media players and streaming services and there was an alternative encryption protocol that was no less efficient but had no exclusionary effects. Should such changes be presumptively lawful? Scholars and courts have proposed two alternative approaches to this question: profit sacrifice, or the no economic sense test, and the rule of reason.

### Profit sacrifice/no economic sense test

Janusz Ordover and Robert Willig suggest an approach to evaluating conduct that has exclusionary effects, including innovation, which does not focus on the distinction between a hard and soft switch.[46] The nub of their approach is the concept of profit sacrifice: Predatory objectives are present if a practice would be unprofitable without the exit that it causes, but profitable with the exit. Conduct that sacrifices profits in order to exclude rivals, and that would not be profitable without that exclusion, is conduct with predatory intent.

Other scholars have built on the notion of profit sacrifice and proposed the no economic sense test (NEST) to identify conduct that falls within the prohibitions of the antitrust laws.[47] NEST advances the principle that conduct that allegedly threatens to create or maintain a monopoly is anticompetitive if, but only if, it makes no business sense for the defendant except for the exclusion of rivals and resulting supracompetitive recoupment. The concept is similar to profit sacrifice, in that a practice that makes no economic sense is a practice that sacrifices profit, but the converse need not hold. Firms can engage in conduct that is marginally less profitable than the conduct that would maximize profits without the exclusion of rivals, but it can still be conduct that makes economic sense. I focus on NEST in the following discussion because it is difficult for antitrust enforcers to determine whether a firm has engaged in conduct that falls short of maximizing profits. Furthermore, innovation typically requires a firm to make costly investments, which can be erroneously labeled as a profit sacrifice.[48]

NEST is an aid to understanding a firm's intent, but it is an imperfect tool to determine antitrust liability, and it is arguably useless to assess innovation that is not accompanied by some type of exclusionary conduct. Successful innovations exclude rivals and many important innovations would not be profitable if they did not exclude rivals. Recall the quote from Steve Jobs at the beginning of chapter 4: "What's the point of focusing on making the product even better when the only company you can take business from is yourself?"[49] Apple invested billions to invent and improve the iPhone. Some of these investments would have made no economic sense if Apple did not expect its iPhone to displace sales of rival mobile phones. More generally, NEST is not applicable to situations in which it is impossible to isolate the benefits from excluding competition from the benefits that are a consequence of legitimate competition on the merits.

NEST has utility in some circumstances. The FTC accused the Intel Corporation of engaging in discriminatory and deceptive conduct intended to maintain its monopoly in central processing units (CPUs) and create a monopoly for itself in graphics processing units (GPUs), including product designs that intentionally defeated interoperability. The FTC alleged that Intel redesigned its compiler and library software to reduce the performance of competing CPUs, pressured independent software vendors not to label their products as compatible with competitors' microprocessor products, even though those products were compatible, and adopted a new policy to deny interoperability for certain competitive GPUs. According to the complaint, many of Intel's design changes to its software had no legitimate technical benefit and were made only to reduce the performance of competing products relative to Intel's products.[50] Although the FTC did not explicitly reference NEST, its allegations reflect a conclusion that Intel engaged in product designs that made no economic sense other than to exclude competition.

The FTC resolved its Intel complaint with a consent decree. In addition to many other obligations and conditions, the decree (which remains in force until 2020 unless modified by the Commission) prohibits Intel from making any engineering or design change to a product if that change degrades the performance of a competitor's CPU or GPU without an actual technical benefit to the product.[51]

NEST is generally conservative, because conduct that harms consumers can pass the test. NEST would not challenge a minor technical improvement to interoperability protocols that excludes competition provided that the improvement has some value and does not incur a disproportionate expense. Similarly, NEST would allow an inexpensive change to a drug that defeats generic substitution if the change would generate a small increase in sales if there were no generic rivals.

NEST is also ambiguous because it does not specify how much exclusion is necessary for the test to apply. Foreclosure is not necessary for conduct to be anticompetitive in markets with strong network effects because network effects amplify the harm from behavior that tips the scales against rivals. Furthermore, although NEST is conservative in many circumstances, it can err by condemning conduct that has consumer value, particularly in markets with excess inertia. In such markets, costly actions (such as defeating interoperability) can promote efficient adoption of a new technology, but may not make economic sense unless they prevent consumers or firms from continuing to use an old technology.

Firms often make expensive bets on new technologies that can displace rivals. These bets may fail to realize expectations and consequently may not make economic sense ex post. That does not mean that the firm's investment decisions were motivated ex ante by the desire to eliminate competition. For these reasons, advocates of NEST argue that it should apply to a firm's expected payoffs, not to realized outcomes,[52] although a firm's expectations are difficult for a court to assess.

Despite these limitations, NEST can provide some insight into whether a firm intends a design change to maintain or extend its dominance for reasons other than competition on the merits. NEST can be a useful tool to identify conduct that has no economic justification, but it is a weak test for conduct that can harm competition. In principle, if not in practice, rule of reason analysis can assess the costs and benefits from product innovations more accurately.

### The rule of reason
The court of appeals in *US v. Microsoft* followed a rule of reason analysis to determine whether product designs by Microsoft were anticompetitive. The court first considered the threshold question of whether the challenged design had an anticompetitive effect. If the court concluded

that the design had an anticompetitive effect, it then investigated whether Microsoft had a procompetitive justification. The plaintiff had the burden to demonstrate anticompetitive effects, and the defendant had the burden to demonstrate procompetitive benefits. If both effects were present, the plaintiff had the burden to show that the anticompetitive effects outweighed the procompetitive benefits.

The rule of reason analysis described in *US v. Microsoft* is appealing in some respects and has been followed by other courts. Unlike shortcuts such as the hard/soft switch dichotomy and NEST, the rule of reason framework is similar to a consumer welfare analysis that fully captures costs and benefits and is favored by some antitrust scholars.[53] But there are two fundamental problems with this framework.

First, the question of whether conduct has an anticompetitive effect should be a determination of the analysis, not a threshold assumption.[54] A product design that disadvantages rivals is not inherently anticompetitive. It is important to focus on the purpose of the analysis and not begin with a presumption that a market outcome or conduct is inherently anticompetitive or procompetitive. Take the case of the camera system developed by Kodak and challenged by the plaintiff in *Berkey v. Kodak*. The new system imposed costs on Berkey and other suppliers of cameras and developing services that did not support the new format. These are costs, but it would be premature to call them anticompetitive effects.

A second problem is the difficulty of balancing alleged anticompetitive effects and procompetitive justifications. That was not an issue in *US v. Microsoft* because, as discussed in chapter 8, the court concluded that the challenged product designs either had anticompetitive effects or procompetitive benefits, but not both. If the evidence is not simply one-sided, courts would have to engage in a quantitative balancing of harms and benefits, which they rarely do.[55] Such an exercise would be particularly difficult in cases of significant innovations or important new product designs. Moreover, a proper analysis of costs and benefits should not end with the short-run effects of innovations or design changes on prices and consumer choices; rather, it should also consider long-run effects, including incentives for future innovations and possible spillover benefits.[56] For some conduct, a rule of reason analysis might identify a less restrictive alternative that has comparable benefits, with less harm to competition. That would require courts to second-guess technological decisions, which they are reluctant to do.

Despite these limitations, the rule of reason is useful to assess product designs that have trivial value and significant exclusionary effects. The peculiar circumstances of the industry for patented pharmaceuticals make this fertile ground to apply the rule of reason to evaluate drug changes. Manufacturers of patented drugs have been accused of "product hopping" (also called "line extensions" or "evergreening") by making minor changes to patented drugs on the eve of patent expiration that, when coupled with removing the old drug or aggressively promoting the new drug, defeat generic substitutions. For these very minor changes (such as a change in packaging from a tablet to a capsule), it would not be particularly difficult for courts to weigh the benefits from these changes against their exclusionary effects for generic competition.

I propose a truncated rule of reason, which would ease the burden on courts to undertake a complex analysis of the costs and benefits of innovations and product designs in most situations. Under this approach, an innovation or product design would be presumptively lawful if it is substantial and not accompanied by exclusionary conduct that is separable from the improved product or technology. Modest innovations or changes in product designs would be candidates for a full rule of reason analysis that compares their benefits to their exclusionary effects. Without regard to whether an innovation or product design is substantial, exclusionary conduct that is not necessary for the innovation or product change would require a rule of reason analysis to determine whether it has benefits that more than compensate for any harm to competition. The truncated rule of reason analysis also can admit consideration of less restrictive alternatives that have similar benefits with less of an exclusionary effect.

In most situations, a modest design change or innovation would not have significant exclusionary effects because an alternative is available that has comparable benefits. There are exceptional circumstances, such as a change in the formulation of a drug that defeats generic substitution or a change to an interoperability protocol. In these cases, the truncated rule of reason can reach a conclusion that the design change or innovation is anticompetitive even if it is not accompanied by other exclusionary conduct. Courts could then require the defendant to abandon the change or make commitments to ensure the availability of comparable alternatives, such as by committing to supply and promote the older formulation of a drug or by providing a way for products to interoperate using an older protocol.

The logic of this proposed truncated rule of reason is that substantial innovations or new product designs have societal benefits that are very difficult for courts to quantify and often exceed the private return to the innovator. Judicial scrutiny of substantial innovations or new product designs risks deterring beneficial investment and would consume administrative resources without corresponding benefits for consumers. These types of innovations or design changes should be shielded from antitrust liability if they are not accompanied by other conduct that has exclusionary effects. Modest innovations or new product designs should not benefit from these absolute protections. By their nature, modest improvements are easier to quantify and in some cases can have exclusionary effects that far outweigh their benefits.

Of course, the threshold for a substantial innovation or new product design is a critical factor in this approach. Whether an innovation or a new product design is a substantial improvement should be assessed by the economic benefit the innovation or product design provides for consumers. Although this can be difficult to quantify, courts are familiar with evaluating complex economic testimony regarding costs and benefits; innovations and product designs are not different in this respect. Furthermore, if an innovation or product design has benefits that are so extensive that they cannot be quantified, that should weigh in favor of a determination that the innovation or product design is substantial.

**How would these principles apply to the Google Shopping case?**
Neither the FTC nor the EC explained its decision in its Google Shopping investigation with specific reference to coercion, NEST, or the rule of reason. The hard/soft switch dichotomy is not very useful to evaluate the Panda update to Google's search algorithms. Like Apple's changes to its FairPlay encryption protocols, the update was inherently a hard switch for rival CSS websites, although the change did not prevent the sponsors of these websites from marketing their services in ways that did not depend on organic search rankings. If Google had procompetitive business reasons to change its display algorithms in a way that demoted rival CSS websites because they made search results more informative, it would be odd to condemn the change as coercion. On the other hand, the Panda update would be coercive without compensating benefits if its only purpose is to suppress competition rather than provide more informative organic search results.

A NEST for the Google Shopping case would have two elements. The first is whether the Panda update had a procompetitive business justification by making organic search results more informative, or was merely an effort by Google to suppress competition by demoting rival CSS websites. The latter would not pass NEST. A second element is the opportunity cost of Google's favorable placement of its Shopping vertical. The allocation of premium SERP real estate to Google's Shopping vertical would not pass NEST if it cost Google more in foregone advertising revenues than Google could expect to earn from its Shopping vertical. The FTC commented on the first element with its brief statement that Google adopted the design change to improve the quality of its search results, but did not comment on the second element, and the EC's decision did not address either element.

The EC did not specifically object to the Panda update or reach a conclusion about its technological merit. More specifically, the EC did not address whether Foundem and other CSS websites should have scored higher in the list of organic search results on the Google SERP. Rather, it objected to the fact that Google was favoring its own Shopping vertical, while demoting rivals:[57]

The Commission Decision does not object to the design of Google's generic search algorithms or to demotions as such, nor to the way that Google displays or organises its search results pages (e.g. the display of a box with comparison shopping results displayed prominently in a rich, attractive format). It objects to the fact that Google has leveraged its market dominance in general internet search into a separate market, comparison shopping. Google abused its market dominance as a search engine to promote its own comparison shopping service in search results, whilst demoting those of rivals. This is not competition on the merits and is illegal under EU antitrust rules.

The FTC did not offer any details that would inform a rule of reason analysis of Google's conduct. The EC made concluding statements about the effects of Google's conduct on prices and innovation that could be a distillation of a rule of reason analysis. However, the EC provided no quantification of the value of Google's design changes and their exclusionary effects to justify its conclusions.

The rule of reason is a useful tool to evaluate some types of innovations, and in particular to allegations of product hopping in the pharmaceutical industry, in which the product innovations are trivial and the exclusionary effects are very large. It is not a useful tool to evaluate complex and potentially valuable design changes in a dynamic industry such as internet search, unless the product designs at issue are clearly trivial changes. For non-trivial product improvements, these considerations bring us back to the standard applied by many US courts to evaluate alleged anticompetitive innovation, which is that legitimate innovation does not violate antitrust laws unless it is accompanied by anticompetitive conduct. A related inquiry is whether there might have been a less restrictive alternative to the Panda update that would accomplish the update's asserted procompetitive benefits without demoting rival CSS websites.

In the Google Shopping case, Google's conduct reduces the incentives of rival CSS to improve their products, but it also encourages them to invest in other services, such as providing manufacturers with assistance to advertise their products on the SERP. At the same time, Google's conduct increases the company's incentive to improve the Google Shopping vertical because it has greater exposure to consumers, and hence greater demand from advertisers. Improvements in the quality of search results benefit advertisers, including the merchants that choose to pay for placement in Google Shopping. These longer-lasting effects confound an already difficult analysis.

Following the truncated rule of reason, the FTC would have reached the correct decision if Google's design changes had substantial consumer benefits and if the Panda update that demoted CSS had a valid business justification, provided that there were no less-restrictive alternatives that would have similar benefits without harming competition. On the other hand, Google's prominent placement for its shopping vertical could have been deemed an exclusionary exercise of market power if the design had no significant cognizable technological merit, the Panda update that demoted rival CSS websites had no valid procompetitive business justification, and CSS is a separate relevant product market in which competition or innovation could be harmed. Antitrust liability would require evidence that the exclusionary effects from the design changes more than offset any benefits. That would require more analysis than the evaluation reported in the decision published by the EC.

## 6   Some Concluding Remarks about Different Antitrust Regimes

Antitrust enforcement for product designs that have exclusionary effects is one of the most difficult challenges for competition policy. The outcomes of the FTC and EC investigations into Google's alleged display bias for product comparison shopping reflect markedly different approaches to this problem. The US has evolved a relatively laissez-faire policy that does not obligate firms to assist their rivals. EU antitrust law has evolved in a different direction, prohibiting discrimination for being an abuse of dominance.

Neither policy is appropriate to adequately address concerns in the high-technology economy. The US policy is too permissive, while the EU policy can deter innovation and punish conduct for which there is no practical remedy. Indeed, while prohibiting dominant firms from discriminating against their rivals is a desirable objective, available remedies often do not achieve that objective without miring courts in complex and costly oversight.

US antitrust law has not always held that firms have no duty to assist their rivals.[58] In 1951, the Supreme Court decided *Lorain Journal v. United States*, a case that involved a refusal by a firm to provide services to customers that patronized a rival.[59] Lorain Journal was the sole publisher of a newspaper in a small midwestern city at a time when newspapers were an important two-sided platform for information and advertising. In response to the arrival of a local radio station, Lorain Journal refused to run ads for advertisers that also ran ads on the radio station. The Supreme Court held that Lorain Journal's refusal to deal with these advertisers was an unlawful attempt to maintain its monopoly power.

The *Lorain Journal* case has a bearing on the Google Shopping case, although there are important factual, legal and economic differences. Lorain Journal conditioned a refusal to deal on its customers' use of a competing medium. The EC did not allege that Google refused to deal with comparison shopping services or discriminated against CSS rivals that advertised in other media. Furthermore, *Lorain Journal* did not involve a new product design and the newspaper offered no plausible efficiency justification for its refusal to accept ads by advertisers that also advertised on the radio station. Nonetheless, both cases allege that a dominant firm sought to exclude competition by discriminating against a rival, either directly by demoting rival search results or indirectly by refusing to deal with a rival's customers.

In *Otter Tail Power v. United States*, the Supreme Court held that the refusal of an electric power utility to transmit wholesale power to communities that established their own retail power systems was an unlawful exercise of monopoly power.[60] This is another case that preceded *Trinko* in which the Court held that discrimination against a rival can be a basis for a finding of unlawful monopolization. The precedents established in *Lorain Journal* and *Otter Tail* do not imply that Google's alleged suppression of rival CSS websites was anticompetitive. However, competition policy would be served by paying attention to the guidance in *Lorain Journal* and *Otter Tail* rather than adhering to a principle that dominant firms have no duty to assist their rivals.

Courts have yet to adopt a uniform and economically defensible approach to evaluate innovation and product changes that exclude competition. Neither the rule of reason analysis, NEST, nor the hard/ soft dichotomy provides a generally useful framework for antitrust analysis of alleged anticompetitive innovation. The truncated rule of reason analysis suggested in this chapter offers a middle ground that defers to significant innovations if they are not accompanied by avoidable exclusionary conduct. For minor product changes and avoidable conduct that has exclusionary effects, the truncated rule of reason requires that antitrust authorities weigh the exclusionary effects against consumer benefits, and may condemn product changes for which exclusionary effects bear a disproportionate relationship to the value of the change.

Antitrust authorities should avoid extreme positions regarding liability for innovations and product designs that exclude rivals, and adopt a middle ground aided by general principles such as the proposed truncated rule of reason approach. It is likely that other antitrust cases that are forthcoming in the high-technology economy will test alternative ways to deal with these difficult issues.

# 10 Competition Policy for Standards

[A] standard-setting organization ... can be rife with opportunities for anti-competitive activity.
—*American Society of Mechanical Engineers v. Hydrolevel Corp.*, US Supreme Court (1982)

## 1 Introduction

Standards are ubiquitous and mostly invisible, but the high-technology economy could not function without them.[1] Generally, there are three types of standards: (1) technical, (2) minimum quality or safety, and (3) informational. Technical standards provide specifications for new and improved products, and they promote economies of scale, competition, and innovation by allowing firms to specialize in compatible components. Quality and safety standards, along with certifications such as the Underwriters Laboratories mark, reduce the risk of buying shoddy or unsafe goods. Informational standards guide the way that facts are reported or work is performed. The focus in this chapter is on technical standards because they are most relevant to competition and innovation in the high-technology economy.[2]

Standards can be developed unilaterally by a single firm that, by virtue of timing or market power, has the ability to impose its will on the market. These include specifications such as Adobe's Portable Document Format (PDF), the x86 microprocessor architecture developed by Intel, and Apple's FairPlay digital rights management technology. They also include metastandards such as the Microsoft Windows, Apple iOS, and Google Android operating systems, which comprise many technologies, some of which are themselves defined by standards.

Alternatively, standards can arise from cooperation between firms and other interested parties. Many established standards development organizations (SDOs) supervise the development of standards at the international and national level. International SDOs include the International Organization for Standards (ISO), the International Telecommunications Union (ITU), and the International Electrotechnical Commission (IEC). Regional, multinational SDOs include the European Telecommunications Standards Institute (ETSI) and the European Committee for Standardization. Some standards are developed with the cooperation of several different multinational SDOs. An example is the third-generation partnership project for mobile telecommunications (3GPP).

In the US, the American National Standards Institute (ANSI) supervises the development of thousands of standards that affect many business sectors. ANSI also evaluates and accredits other organizations for conformance with its standard-setting guidelines. They include the Institute of Electrical and Electronic Engineers Standards Association (IEEE-SA); the Joint Electron Device Engineering Council (JEDEC); the Motion Picture Experts Group (MPEG), which sets standards for audio and video compression and transmission; and ASTM (formerly the American Society for Testing and Materials), which sets standards for a wide range of materials, products, systems, and services. The National Institute of Standards and Technology (NIST), established in 1901 to improve measurement services, develops standards in a wide range of industries, from smart electric power grids to nanomaterials.

Parties that wish to develop standards can bypass established SDOs and form standard-setting consortia (also called "special interest groups"), which can range from ad hoc groups to structured organizations that may or may not have ANSI accreditation. Examples include the Internet Engineering Task Force (IETF) and the World Wide Web Consortium (W3C), which provide forums for network designers, operators, vendors, and researchers concerned with the evolution and smooth operation of the internet; the Blu-Ray Disc Association, for high capacity optical disk storage; and the universal serial bus (USB) for computer connections, originally developed by a group of seven companies and now managed by the USB Implementers Forum.

Most standards that are developed unilaterally or cooperatively are not legally binding on manufacturers or firms that use products related to the standards. Some standards are promulgated by law, and compliance is enforced by agencies such as the Federal Communications Commission or the Environmental Protection Agency.

Whether a standard is open can have significant policy implications, but the term has no standard definition. Some refer to standards as "open" if they are developed following ANSI guidelines, which emphasize inclusiveness, transparency, and consensus. Others use the term to mean that the standard is available to anyone, not confined to particular firms or business applications. Still others use "open" to mean that the standard is not encumbered by proprietary intellectual property rights, which is the definition employed in this chapter.

A proprietary standard is owned by a firm or a group of firms, or covers technologies that have patent protection or software protected by copyright. Firms that own the rights to a standard may contractually restrict its use. An example is the Dolby Digital standard for audio compression, which Dolby licenses for use by manufacturers of audio and video equipment. Depending on the rules that govern the development of a standard, firms that own patents that cover technologies specified in a standard may charge royalties or refuse to license their patent rights, even if the standard was developed in an open process.

Standards vary along a continuum between truly open and proprietary. Many standards are proprietary but licensed without charge. A standard can be open for use by anyone without compensation, but it may require users to comply with conditions such as not making unauthorized changes to the standard or not exerting property rights that cover the standard. Other standards are open, but users have added features that have intellectual property protection. The Linux operating system kernel (the software that interfaces with the computer hardware) is "open source," which means that the Linux source code is freely available and may be redistributed and modified, although a number of extensions that are proprietary have been added to the Linux kernel.

Firms that promote a particular standard have to convince consumers and the suppliers of complementary products and services to support their preferred standard. These so-called standards wars can incur wasteful duplication and impose costs on firms and consumers that have made commitments to a defeated and incompatible technology.[3] Examples of standards wars include the struggle between the incompatible VHS and Betamax videotape recording formats in the late 1970s and 1980s, and more recently, between the incompatible Blu-ray and High-Definition Digital Versatile Disk (HD-DVD) optical disk standards.

Cooperative efforts to develop standards can avoid standards wars that occur when firms promote different preferred standards, but cooperation can be costly and does not always produce the best outcome. Members of an SDO or consortium may fail to agree on a standard or can take years to reach consensus.[4] The process of developing a new standard within an SDO can become corrupted by special interests, resulting in standards being chosen that do not provide the most economic value.[5]

Standards often create consumer benefits and promote competition and innovation, but standards also can raise concerns that warrant antitrust vigilance.[6] Section 2 addresses several types of unilateral conduct associated with standard-setting that have raised antitrust concerns, including conduct related to the exercise of intellectual property rights. Section 3 addresses antitrust issues for coordinated conduct in standard-setting. Standard-setting often brings together firms that compete in markets that would implement the standard. This naturally raises the possibility that the firms can conspire to set standards that are mutually profitable, but not in the best interests of consumers. They can also use the standard-setting venue to adopt intellectual property policies that promote the interests of the participants in the standards development process but do not promote innovation or benefit consumers. Section 4 offers proposals related to commitments to license patents that are essential to practice a standard that can have benefits for competition and innovation.

## 2   Antitrust Issues for Unilateral Standard-Setting

Antitrust concerns for conduct by a single firm generally involve attempts to maintain or achieve monopoly power. Standards have been involved in such attempts in at least three ways:

1. Efforts to disrupt industrywide compatibility with a standard

2. Efforts to compel adherence to a dominant firm's preferred standard

3. Efforts to capture value from a standard by exerting proprietary intellectual property rights

### Efforts to disrupt industrywide compatibility with a standard

The Microsoft case discussed in chapter 8 illustrates possible competitive effects when a dominant firm fragments an industry standard to promote its own customized version of that standard. Microsoft licensed the Java technologies from Sun Microsystems. A standard Java implementation promised easy porting of computer applications across different operating systems, which could break the applications barrier to entry that supported Microsoft's personal computer (PC) operating system monopoly.[7] The US Department of Justice (DOJ) and the plaintiff states alleged that Microsoft's efforts to splinter the Java standard were a violation of the Sherman Act.

The district court held that Microsoft's effort to splinter the standard violated the antitrust laws. The court of appeals affirmed this finding, but it did not require Microsoft to conform to a single Java standard. As noted in chapter 8, a requirement that Microsoft support a common Java implementation has an economic justification because the adoption of a Java standard was central to the allegations in the case. The court could have compelled Microsoft to support a common Java standard, while also allowing Microsoft to pursue a proprietary Java implementation. The European Commission (EC) did something similar when it required Microsoft to offer a "ballot box" on the Windows start-up screen, which allowed consumers to choose their preferred browser.

### Efforts to compel adherence to a dominant firm's preferred standard

The Android case pursued by the EC against Google illustrates the second type of standards-related conduct: efforts by a dominant firm to enforce a standard. Google offers a no-fee license for its Android mobile operating system, provided that smartphone licensees also install a collection of Google apps called Google Mobile Services, which include the Google Play store and the Google Chrome browser. Google's royalty-free license prohibits manufacturers from selling devices that run on a different version of Android (known as a standard "fork"). The EC held that Google could not prevent developers from developing Android forks by denying them licenses to the Google Mobile Services collection of mobile apps. In its decision, the EC stated that "Android forks constitute a credible competitive threat to Google" by allowing smartphone developers to develop differentiated versions of Google's Android operating system at relatively low cost.[8] The EC found that Google's licensing policies contributed to an abuse of dominance by tying Google apps to the operating system[9] and, in July 2018, it levied a fine of 4.34 billion euros.[10]

The EC rejected Google's argument that the licensing restrictions were necessary to prevent fragmentation of the Android ecosystem. Instead, it argued that Google's refusal to authorize forks of its Android operating system impeded innovation and competition from smart mobile devices based on alternative versions of the Android operating system.

Allowing a standard to splinter into different specifications undermines the power of the standard to promote compatibility, but it also allows greater product diversity and innovation. Google's Android mobile operating system is based on a modified version of the Linux kernel. Apple's iOS mobile operating system is based on a version of Unix, which is a predecessor to Linux. If one had to choose between a single standard version of Unix or Linux and allowing Unix or Linux to fragment, the latter was the better choice because it facilitated innovation and competition in mobile operating systems.

### Efforts to capture value from a standard by exerting proprietary intellectual property rights

No aspect of standards-related conduct has attracted more scrutiny (at least as measured by the number of investigations and policy papers) than conduct related to the disclosure and exercise of standards-essential patents (SEPs). Most SDOs allow standards to include proprietary intellectual property rights and do not prohibit rights owners from charging royalties. The ANSI patent policy states, "There is no objection in principle to drafting an American National Standard (ANS) in terms that include the use of an essential patent claim (one whose use would be required for compliance with that standard) if it is considered that technical reasons justify this approach."[11] Some standards specify technologies that are covered by hundreds of patents. One study identified over 700 patent families that were declared essential to the WCDMA cellular standard and over 500 patent families declared essential to the CDMA2000 cellular standard as of early 2004.[12]

The development of standards that are covered by propriety intellectual property rights often pits two camps against each other: innovators, who own patent rights and want compensation for use of their rights; and implementers (who can also be innovators for follow-on discoveries), who sell products that implement the standard and want a low cost for the standardized technologies. Implementers complain about the risk of holdup by patentees who charge high royalties for patents on a standard after they make investments that lock themselves into the use of the standard.[13] Innovators, on the other hand, complain about the risk of holdout by implementers, who want to practice their patents without fair compensation for their research and development (R&D) expenditures. Both complaints allege opportunistic behavior to exploit sunk costs. For implementers, it is the cost of investments to develop products that comply with the standard, and which cannot be repurposed to other products. For innovators, it is the cost of R&D that was necessary to make the patented discoveries.

Patent rights can interfere with innovation and allow patent owners to impose high royalties when products embody numerous technologies with many owners. Each patent owner that is active in this patent thicket has an individual incentive to demand a large share of the value of the product, and the resulting total royalty demand can exceed the royalties that would maximize a licensor's profit if it were the sole source for all the patents.[14] Implementers complain about this "royalty stacking" that results from demands by separate patent owners and fear injunctions that prohibit sales of products that infringe a patent.[15] Standards aggravate concerns from injunctions and royalty stacking if firms and consumers make investments that are specific to the standard and face high costs to switch to an alternative standard.

Many SDOs address the conflict between implementers and innovators by requiring, prior to the approval of a proposed standard, that innovators declare patent rights that they own (and sometimes patent applications) that are essential to the standard specification, and to agree to license such patents at terms that are fair, reasonable, and nondiscriminatory (FRAND, also called RAND or F/RAND). The ANSI patent policy states:[16]

The [ANSI-Accredited Standards Developer] shall receive from the patent holder or a party authorized to make assurances on its behalf, in written or electronic form, either:

a) assurance in the form of a general disclaimer to the effect that such party does not hold and does not currently intend holding any essential patent claim(s); or

b) assurance that a license to such essential patent claim(s) will be made available to applicants desiring to utilize the license for the purpose of implementing the standard either:

i) under reasonable terms and conditions that are demonstrably free of any unfair discrimination; or

ii) without compensation and under reasonable terms and conditions that are demonstrably free of any unfair discrimination.

Unfortunately, SDOs have not defined the limits on FRAND terms. Furthermore, they do not have uniform disclosure requirements or uniform definitions of "essential." Studies show that many patents declared essential to common standards are not technically nor economically necessary to implement the standard.[17]

Antitrust authorities in the US and elsewhere have challenged several alleged failures of SEP owners to disclose their patents or license them on FRAND terms. One of the first of these cases involved Dell Computer and the VL computer bus standard for communicating data between a central processing unit and peripheral devices. The VESA SDO, of which Dell was a member, standardized the VL bus design in 1992. Before VESA approved the standard, a Dell representative certified that the proposed standard did not infringe any of Dell's trademarks, copyrights, or patents. However, Dell had received a patent on the VL bus design a year earlier and sought to enforce its patent after computer manufacturers adopted the standard. The Federal Trade Commission (FTC) lodged a complaint, alleging that VESA would have implemented a different nonproprietary design had it been informed of the patent conflict during the certification process.[18] The Commission and Dell reached a settlement, according to which Dell agreed not to enforce its patent against anyone implementing the VL bus standard.[19]

The FTC pursued its concerns about the disclosure of SEPs in several other cases, including extensive litigation against Rambus, a company that develops and licenses technologies for computer memory and other devices.[20] The allegations involved Rambus's conduct as a participant in JEDEC, an SDO that develops standards for, among other devices, dynamic random access memory devices (DRAMs).[21] Rambus participated in JEDEC for a period of time, during which it did not disclose that it had patent applications and plans to apply for patents that covered standards being developed for DRAMs, including for synchronous dynamic random access memory (SDRAM) devices. After JEDEC issued SDRAM standards, Rambus demanded royalties for infringement of patents it owned that covered the standards. The FTC alleged that Rambus's silence and subsequent enforcement of its patents allowed Rambus to monopolize key technology markets for these devices.

Rambus appealed. The Court of Appeals for the DC Circuit (the same court that issued the final ruling in *US v. Microsoft*) ruled in Rambus's favor.[22] The gravamen of the court's decision was the failure of the FTC to prove that JEDEC would have chosen a *different* standard that did not infringe Rambus's patents if these patents had been disclosed. The court stated:[23]

[The Commissions'] factual conclusion was that Rambus's alleged deception enabled it *either* to acquire a monopoly through the standardization of its patented technologies rather than possible alternatives, *or* to avoid limits on its patent licensing fees that the SSO would have imposed as part of its normal process of standardizing patented technologies. But the latter—deceit merely enabling a monopolist to charge higher prices than it otherwise could have charged—would not in itself constitute monopolization.

The court observed that JEDEC's patent policies were muddled during the time that Rambus was a member, and did not clearly require the disclosure of planned applications for patents that may be necessary to make or use standard-compliant products. Nonetheless, the court held that had Rambus engaged in deceptive conduct, it would not have been a violation of the Sherman Act if it did not affect JEDEC's chosen standards.[24] The court emphasized this conclusion, notwithstanding that JEDEC's patent policy would have required a commitment from Rambus to license its patents at FRAND terms as a condition to approve the standards, which likely would have resulted in lower negotiated patent royalties.

The court's reasoning can be understood only in the context of arcane US antitrust law. US antitrust law prohibits monopolization; it does not prohibit high prices. The court distinguished its conclusion from that of *US v. Microsoft*, where the same court held that Microsoft's deceptive claims about the cross-platform compatibility of its Java tools was a violation of Section 2 of the Sherman Act (but nonetheless did not compel Microsoft to support an industry-standard Java implementation). The distinction is that Microsoft's conduct excluded competition and contributed to the maintenance of its Windows monopoly, whereas Rambus's conduct arguably only resulted in higher prices. With regard to Rambus, the court emphasized, "Indeed, had JEDEC limited Rambus to reasonable royalties and required it to provide licenses on a nondiscriminatory basis, we would expect *less* competition from alternative technologies, not more; high prices and constrained output tend to attract competitors, not to repel them."[25]

Conduct that enables a patent owner to evade FRAND commitments should not be lawful. High royalties harm consumers and can impede innovation for technologies for which a patent license is necessary. Some have argued that patent holdup is no more than an academic curiosity because innovation and competition for smartphones and other devices have thrived, despite the fact that these devices implement standards covered by hundreds of SEPs.[26] But this argument is flawed. It does not recognize that prices for smartphones and other devices would likely be much higher if the antitrust authorities and the courts stopped policing FRAND licensing obligations.[27] The fact that it is reasonably safe to drive on highways in the US does not mean that speed limits are unnecessary. FRAND limitations are speed limits on the information superhighway.

Organizations that develop standards are joint ventures whose members include firms that are actual or potential competitors. Joint ventures can benefit consumers by creating new or more efficient products or by creating standards that facilitate new or more efficient products. They can also harm consumers by serving as venues to fix prices, reduce quality, raise costs, or exclude competitors. As the next section shows, SDOs are not immune to these concerns.

## 3   Antitrust Concerns from Collective Conduct by SDOs

Collective conduct by SDOs and their members has raised antitrust concerns for a variety of reasons. In the high-tech sector of the economy, these concerns mostly involve collusion to exclude competing technologies and policies that SDOs adopt to address royalties and other licensing terms for SEPs. SDOs are ripe for collusion because they bring together actual or potential competitors with powerful commercial interests that can bias standards choices. Antitrust enforcers have been understandably reluctant to second-guess technological choices made by SDOs and their members, but they have not hesitated to challenge abuse of the standards development process that allegedly has anti-competitive consequences.[28]

An example of alleged abuse of the standard-setting process is "vote stacking," in which members of a standards body recruit participants to vote affirmatively for their interests. The Supreme Court addressed vote stacking by an SDO in *Allied Tube & Conduit Corp. v. Indian Head*, which involved standards for conduit to protect electrical wiring. The National Fire Prevention Association (NFPA) publishes the National Electrical Code, which establishes requirements for the design and installation of electrical wiring systems. Indian Head, a manufacturer of plastic conduit, submitted a proposal to extend the code to approve the use of plastic, as well as the conventional steel, conduit.

Allied, the largest producer of steel conduit in the US, colluded with members of the steel industry, other steel conduit manufacturers, and independent sales agents to recruit new NFPA members whose only function was to vote against the proposal to approve the use of plastic conduit. Their recruiting efforts were successful, and the proposal was defeated. In response, Indian Head sued, alleging that Allied and others had unreasonably restrained trade in the electrical conduit market, in violation of Section 1 of the Sherman Act. After a number of decisions and reversals, the Supreme Court sustained a verdict for Indian Head, stating, "What [Allied] may not do (without exposing itself to possible antitrust liability for direct injuries) is bias the process by, as in this case, stacking the private standard-setting body with decisionmakers sharing their economic interest in restraining competition."[29]

The court did not define vote stacking or limit the circumstances in which it should raise antitrust concerns. Participation in standards-setting meetings is driven by economic interests, and firms with more at stake in the standards process tend to supply more participants to relevant standard-setting organizations.[30] Vote stacking could be defined as recruiting participants who are not employees of interested firms, but that would not prevent firms that can tap a large labor pool from dominating the standard-setting process.

It is not clear that vote stacking results in inferior economic outcomes compared to more limited representation in a standard-setting organization. Unlike the outcome of market competition, which generally benefits consumers by resulting in lower prices, there is no corresponding "invisible hand" principle that standard-setting bodies will coordinate on the best standard, even if the process of standard-setting is open and transparent. Kenneth Arrow (who proved central results for innovation competition, as described in chapter 3) established this result as a corollary of his famous "impossibility theorem," which proves that voting rules (such as those followed by SDOs) cannot generally assure outcomes that satisfy reasonable conditions for rational choice.[31]

Nonetheless, allegations of anticompetitive exclusion by SDOs have typically failed when courts concluded that the standards were developed following an open and transparent process, when participation in standard-setting was unrestricted and nondiscriminatory, and when compliance with the standard was voluntary.[32] Courts have also considered allegations of anticompetitive standard-setting when these conditions were not satisfied. In at least one case, a court refused to dismiss an allegation that a standard-setting consortium harmed competition by engaging in a closed standard-setting process that was intended to give its members a time-to-market competitive advantage over rival manufacturers.[33]

### Coordination in the determination of FRAND patent-licensing obligations

Most SDOs require that patent owners agree to license patents that cover technologies specified in a standard as either royalty-free or at FRAND terms. But FRAND obligations are ambiguous, and FRAND licensing commitments have not resolved tensions between patent owners and the firms that sell products that implement standards covered by the patents.

Several courts have addressed compliance with FRAND licensing and found that owners of FRAND-encumbered patents made royalty demands that greatly exceeded the value of the licensed patents.[34] Some SDOs have attempted to address these concerns by providing opportunities to clarify FRAND obligations ex ante, before the standards issue.

VITA exemplifies an SDO with a strong ex ante patent policy. VITA began in 1982 as the VMEbus Manufacturers Group to create a standard for the Versa Module Europa computer bus. Since 2006, the VITA standards organization (VSO) has required members of its working groups to disclose the existence of all patents and patent applications owned, controlled, or licensed by the member company that the member believes contain claims that may become essential to a draft specification, and to declare the maximum royalty rate and most restrictive licensing terms for these patent claims. Under the VSO policy, failure to disclose an essential patent claim in a timely manner obligates the owner to license the patent on a royalty-free basis.[35]

VITA asked the DOJ for a Business Review Letter (BRL) prior to publishing its VSO patent policy. A BRL is an advisory statement from the Department about its current enforcement intentions regarding a proposed policy or combination. VITA justified its patent policy proposal by citing past experience, in which owners of patents on VSO standards demanded high royalties after the standard issued, which raised costs and delayed the standard's market adoption, and in one instance rendered a proposed standard commercially infeasible. In response, the DOJ recognized efficiencies from VSO's proposed policy by allowing working groups to make more informed decisions about the costs as well as the purely technical merits of alternative standards, while acknowledging that the collaborative standard-setting process could result in exclusionary and collusive practices that harm competition and violate antitrust laws. Applying a rule of reason analysis to potential costs and benefits, the DOJ concluded that it had no intention to challenge the VSO patent policy.[36] This was about as favorable a conclusion as one can expect from a BRL.

IEEE-SA, the SDO for the IEEE, has a much larger portfolio of standards development activities. The IEEE-SA updated its patent policy in 2015 to clarify the meaning of a FRAND offer and to impose other conditions on the licensing of SEPs by members that agree to abide by the licensing commitments. The DOJ reviewed these changes and concluded that the process leading to the modified policy was open and did not raise antitrust concerns. The BRL noted that the modified policy had the potential to improve the IEEE-SA standard-setting process by possibly reducing patent litigation and mitigating holdup and had the potential to benefit competition and consumers by creating greater clarity.[37] The letter also concluded that the DOJ had no intent to challenge the policy.

Notwithstanding the generally favorable BRL responses from the DOJ, the VSO and IEEE-SA policies raise several potential types of antitrust concerns:

• Members of standard-setting committees that make, use, or sell products that implement standards may use the mandatory disclosure of patent-licensing terms as a means to collusively exercise monopsony power to depress royalties below the levels that the implementers can obtain in bilateral negotiations. SDO patent policies disavow such collusion, but that does not mean it cannot happen.

• Members of standard-setting committees that own patents may use the mandatory disclosure of patent-licensing terms as a means to collusively exercise monopoly power to raise royalties above the levels that the implementers can obtain in bilateral negotiations.

• Onerous patent disclosure policies and licensing rules can cause patent owners to choose different standard-setting venues that have more favorable patent disclosure and licensing policies or cause them to withdraw from the activity of developing draft standards.[38]

• Negotiations required by disclosure and licensing rules can delay the progress of standards development.

• Mandatory disclosure and licensing rules can force owners of patents to establish royalties for patents that they would otherwise choose not to assert.

• Requirements to disclose maximum royalties and most restrictive licensing terms can incentivize patent owners to declare artificially high royalties and restrictive terms.

• Mandatory disclosure and licensing rules may cause patent owners to refuse to disclose possible SEPs or refuse to license SEPs at FRAND terms unless required to do so by the SDO.

Some of these concerns are either not likely to occur or have offsetting potential benefits. VSO's ex ante disclosure policy is mandatory but not likely to result in a significant exercise of monopsony power because VITA has long had a policy of promoting open standards. Disclosure delays in the standards development process caused by introducing an additional dimension for negotiation can have procompetitive benefits by allowing implementers to evaluate the economic costs as well as the technical merits of alternative proposals for standards.

## 4  Some Suggested Policies to Promote FRAND Licensing

Antitrust authorities can influence the development of standards and encourage policies for SEPs that protect consumers and promote innovation. They can condition the approval of mergers or resolve allegations of anticompetitive conduct on agreements to support industrywide standards. Alternatively, they can address restrictive conduct by firms that control a standard with conditions that allow firms to develop alternative standards, as the EC has done for the licensing of Google's Android mobile operating system.

With regard to SEPs, courts could expand their concept of monopolization to condemn conduct that increases prices by abusing the standard-setting process instead of limiting liability to conduct that harms a narrow definition of competition. Post-Rambus, most challenges to the abuse of FRAND commitments have alleged a violation of a contract between the patent owner and the licensee rather than an antitrust violation. Antitrust enforcement is a valuable tool for FRAND compliance because, unlike contract enforcement, the plaintiff does not have to be a direct party to the contract. Furthermore, antitrust enforcement can allow remedies, such as compulsory licensing, that may not be available under the law of contracts.[39]

Recent statements by the Assistant Attorney General for Antitrust at the DOJ have emphasized the risk of holdout as well as holdup for the licensing of SEPs.[40] Concerns about holdout are valid, but patent owners have legal recourse to collect damages from technology users that infringe their patents without adequate compensation. Holdup is different in this regard because neither patent, contract, nor antitrust laws clearly define the circumstances in which a licensee is entitled to compensation for licensing demands that violate FRAND obligations. Although courts that have reviewed complaints of FRAND violations did not always find evidence of holdup and royalty-stacking, they often found that SEP owners made royalty demands that exceeded the economic contribution of the patents to the value of the standard. Furthermore, some owners of FRAND-encumbered patents have attempted to avoid FRAND commitments altogether by transferring ownership of the patents to third parties that did not commit to license their patents at FRAND terms.[41] Such avoidance should not be permitted.

Patent pools address the problem of royalty-stacking by offering licenses for the pool's entire patent portfolio.[42] Organizations that develop standards could condition the certification of a standard on agreement by holders of SEPs to join a patent pool that offers a portfolio license. Some special-interest groups, such as Bluetooth and developers of standards for optical disks, offer a portfolio license for their members' patents. However, standards that have hundreds of SEPs owned by parties with different interests are unlikely to achieve broad participation by the patent owners in a portfolio license.

Nonetheless, patent pools have a useful lesson for FRAND compliance. Pools typically publish terms for a portfolio license. The terms may have a fixed fee, a fee that varies with the number of licensed units or sales of downstream products, or a combination of the two, but they are transparent and the same for every potential licensee. In contrast, although SDOs request that SEP owners offer licenses on nondiscriminatory terms, they do not specify what "nondiscrimination" means and SEP owners often negotiate different deals with licensees. Moreover, the deals are often protected by nonconfidentiality agreements, which makes compliance with nondiscrimination difficult or impossible to verify.

A requirement that owners of SEPs publish their terms for licenses would address the "nondiscrimination" prong of the FRAND commitment and help to promote license terms that are fair and reasonable.[43] An additional benefit from a uniform and transparent publication of SEP license terms is that the terms can prevent holdup if they are published before firms and consumers make investments that are specific to a standard. Suppose that a patent owner negotiates a bilateral license with an implementer before a standard has been certified. A nondiscrimination agreement would obligate the patent owner to offer the same terms to licensees after the standard issues. A public nondiscrimination commitment would allow patent licenses that are negotiated ex ante to protect against holdup that could occur after firms and consumers make investments that are specific to a standard.

A public nondiscrimination commitment is not perfect, though. It would discourage patent owners from negotiating better deals to entice reluctant implementers because the patent owner would have to offer the better deal to all licensees. A nondiscrimination requirement might also complicate the ability of patent owners to adjust royalty and licensing terms in response to changing patent values. Nonetheless, transparent and nondiscriminatory licensing terms have benefited both the owners of patents that participate in patent pools and their licensees. A similar arrangement would have advantages over existing confidential FRAND agreements, which make it difficult to enforce reasonable licensing terms and assess compliance with nondiscrimination.

Sensible rules for patent damages can ensure that SEP license terms are fair and reasonable. Patent owners negotiate licensing terms in the shadow of infringement damages. Unfortunately, the law of patent damages has sometimes allowed patent owners to receive compensation for infringement that is unrelated to the economic contribution of the patented technologies.[44] Damages for infringement of a patented technology should reflect the economic contribution of the patented technology, not the value of the standard for which it is a component or an elevated royalty that a patent owner could demand because an implementer of the patented technology would have a high cost to switch to a technology that does not infringe the patent.[45] Patent damages should apportion the value of a standard to infringed patents that cover components of the standard. If 100 patents are clearly essential to practice a standard and there are no substitutes for any of the patents, then each patent accounts for 1 percent of the economic value of the standard. It is encouraging that courts have recently concluded that FRAND royalties should measure the economic contribution of the patented technologies, not the value added by the standard's adoption of the patented technology, and have endorsed the logic of apportionment for patent values.[46]

A patent owner has more bargaining power if a court would approve a demand for an injunction or exclusion order that prevents the use of the patent. Reasonable rules for permitting injunctions and exclusion orders would limit their use to situations in which a potential licensee refuses to negotiate a license and monetary damages cannot adequately compensate the patent owner for infringement.[47]

Robert Merges and Jeffrey Kuhn suggest the concept of "standards estoppel" to prevent attempts by patent holders to hold up licensees by capturing part of the cost of switching to a different standard.[48] Under this principle, the intentional non-assertion of a patent in the presence of its widespread adoption would create immunity from patent infringement. This would prevent the patent owner from delaying assertion to benefit strategically from irreversible investments that lock firms and consumers into compliance with a standard. The Rambus case discussed in this chapter would have been a candidate for their proposal.

# 11 Some Concluding Remarks on Innovation-Centric Competition Policy

As a society, we are conflicted about the digital economy. We bemoan our loss of privacy and worry that the winds of creative destruction that topple dominant firms appear to have slowed. At the same time, we cannot ignore the many ways that the giants of the digital economy have improved our lives with new products and services. Competition policy has to navigate the conflicting costs and benefits of market power in the digital economy. For some, the solution is to abandon the traditional focus of antitrust enforcement on consumer welfare and incorporate broader concerns, such as jobs, privacy, inequality, and the concentration of political power. Including these ill-defined goals increases the risk that courts and antitrust agencies will have too much discretion to respond to political pressures, corporate lobbying, and personal biases. The commitment of antitrust enforcers to policies that promote consumer welfare goes far to explain the relatively weak impact of these influences.

There is another path to achieve broader goals while preserving a focus on consumer welfare. Antitrust enforcement should evolve from being *price-centric* to *innovation-centric*. Price-centric antitrust enforcement prevents mergers that are likely to raise prices, and prevents firm conduct that excludes competition for existing products and services. Innovation-centric antitrust enforcement does not abandon these concerns, but it augments them by challenging mergers and firm conduct that are likely to harm innovation and competition for products that do not presently exist. Innovation-centric competition policy will achieve goals that price-centric enforcement neglects, such as ensuring opportunities for entrepreneurs to compete and thrive.

Although the antitrust laws do not have to be rewritten to address innovation, there are historical precedents that erect judicial barriers to the enforcement of dynamic competition. I summarize some of the most significant obstacles and how they must change to accommodate innovation-centric competition policy.

### Antitrust authorities should not emphasize market definition to analyze innovation and future price competition

Courts have developed analytical tools such as market definition to evaluate antitrust allegations, and they have increasingly demanded empirical evidence to prove anticompetitive effects, which they typically measure by firm shares, prices, and output in existing markets. This policy evolution has introduced commendable analytical rigor to antitrust enforcement. It restrains enforcement that would deter beneficial conduct, and limits the ability of special interests to influence judicial outcomes. Yet this focus of modern antitrust enforcement on market shares and prices also raises barriers to evaluate allegations of harm to innovation and future price competition.

Harm to innovation requires an analysis of effects on research and development (R&D) incentives. Conventional approaches to market definition are generally unhelpful for this analysis because most R&D is not traded in a market. Conventional approaches to market definition also are unhelpful for innovation-centric antitrust policy because mergers and other conduct involving high-tech firms affect the availability and prices of products that may appear in the future, for which the lack of empirical evidence makes the definition of these future markets inherently uncertain.

In contrast to conventional market definition, the concept of a "research and development market" can be useful to analyze innovation incentives and future price competition by identifying the firms with the specialized assets necessary to conduct R&D devoted to a particular application. A merger in an industry with few firms that possess these specialized R&D assets raises concerns about downward pressure on innovation incentives. A merger in an industry with many firms that possess these specialized assets is unlikely to harm innovation or future price competition. Thus, definition of a research and development market can be a useful screen to identify the potential for harm to innovation and price competition, much as conventional market definition is a useful screen to identify the potential of a merger to raise prices for existing products and services.

### Antitrust authorities should rely on validated presumptions to assess innovation effects

Innovation-centric antitrust enforcement will require courts to rely more on presumptions about future effects that are buttressed by economic theory, empirical evidence, and corporate records and testimony. Although available theory makes diverse predictions about the effects of competition and mergers on innovation, it also provides robust insights into different market circumstances. Mergers decrease innovation incentives through a unilateral effect that operates in much the same way that mergers increase prices. Profits that are at risk from innovation can be a heavy drag on innovation incentives. Mergers can stifle innovation even if the merging firms do not supply products that are substitutes for each other.

Courts should rely on an extensive body of theory and empirical studies to support presumptive effects of mergers on innovation and future price competition. Mergers can harm innovation by creating downward innovation pressure, or they can promote innovation if they allow the merging parties to appropriate greater value from their innovations than they could obtain as independent firms. Unfortunately, empirical studies of innovation effects from mergers are few in number; there is great need for additional empirical research to inform competition policy.

### Courts should require evidence of benefits from proposed mergers or acquisitions

The Clayton Act prohibits mergers or acquisitions that are likely to substantially lessen competition in a relevant market. Antitrust enforcers have the burden to prove that harm from a merger is likely and substantial. However, many transactions in the high-tech sector have anticompetitive effects that are substantial, but often difficult to prove with a high degree a certainty. At the same time, many of these transactions have questionable efficiency benefits.

Competition policy would be more effective for the high-technology economy if courts required evidence of benefits from proposed mergers or acquisitions, or from other conduct that has the potential to harm competition and innovation. Existing competition policies prioritize the risk of overenforcement, perhaps based on a belief that market forces can correct anticompetitive conduct, but are less able to correct misguided antitrust intervention. In the high-technology economy, the potential consumer harm from underenforcement of the antitrust laws is at least as great as the potential harm from overenforcement. Forces such as network effects and economies of scale reinforce market dominance and aggravate concerns about market power obtained by excluding rivals or by acquiring potential competitors when the conduct does not have offsetting benefits. Reasonable presumptions should strike a balance between errors of overenforcement and underenforcement to foster innovation and future competition.

**Courts should increase scrutiny of acquisitions that eliminate potential competitors**

The court of appeals in *US v. Microsoft* affirmed the principle that the antitrust laws apply to conduct by a dominant firm that eliminates nascent threats. Exclusionary conduct by a dominant firm can violate the antitrust laws without having to prove that the excluded competition likely would have occurred without the conduct. This principle should also apply to acquisitions of nascent competitors. Dominant firms in the high-technology economy are adept at identifying competitive threats and can acquire them in their infancy, before their targets achieve a market presence that would trigger conventional antitrust concerns.

Many of these potentially harmful acquisitions fall below the Hart-Scott-Rodino (HSR) value thresholds that require reporting of the transaction to the Federal Trade Commission (FTC). Although transactions that fall below the HSR thresholds are not immune from antitrust enforcement, reporting greatly facilitates the detection of troublesome transactions. The HSR thresholds should be modified to require reporting of acquisition targets with modest revenues if the acquirer is a firm that dominates an industry.

Firms in the high-tech sector of the economy have made hundreds of acquisitions and they will make many more. Even if the probability is small that any single acquisition will eliminate a significant competitive threat, there is a high probability that at least one acquisition will have this effect. Courts should not presume that potential competition must be likely and significant to warrant antitrust intervention, and they should oppose acquisitions that have no credible efficiency benefits.

However, courts and antitrust authorities should not presume that acquisitions of potential rivals are anticompetitive if the expectation of acquisition by an established firm in a related technology field is the motivator for innovation by the acquired firms in the first place, provided that there are no other acquirers that would offer similar rewards without the risk of anticompetitive effects. These other acquirers would be less restrictive than a potentially harmful merger. Courts should erect a high bar to acquisitions that may harm competition or innovation if there are other acquirers for which anticompetitive harm is much less likely.

### Courts should not require evidence of substantial foreclosure to prevent exclusionary conduct

Innovation-centric competition policy should continue to emphasize restrictions on conduct that excludes competition and allows firms to benefit in ways that are not the result of superior performance or efficiency. The prevention of conduct that excludes competition is particularly important in markets with network effects because the exclusion of rivals can allow these markets to tip to a dominant supplier that does not offer consumers the best products and services. Antitrust policy should apply a lower threshold than substantial foreclosure for anticompetitive exclusionary conduct in industries with network effects.

### Compulsory licensing is often an effective tool to promote innovation

Most observers of antitrust enforcement would put the breakups of Standard Oil and AT&T at the top of the list of greatest antitrust "hits." But they might overlook compulsory licensing agreements that had profoundly beneficial impacts on competition and innovation. These include the 1956 AT&T and IBM consent decrees, which compelled the licensing of more than 9,000 patents (some royalty-free), as well as the 1975 Xerox consent decree, which compelled the company to offer patents covering plain-paper copiers at reasonable royalties. These decrees were powerful stimulants for competition and promoted follow-on innovations that built on the patented technologies.

Compulsory licensing obligations in merger consent decrees appear to have promoted patenting by firms that were the beneficiaries of these compulsory licenses without significantly reducing patenting by the firms that were compelled to license intellectual property or by other industry participants. These observations are consistent with studies of compulsory licensing in other industry contexts. Compulsory licensing can promote interoperability to benefit innovation by firms that offer complementary products and services. Where data is a barrier to entry, compulsory licensing can offer rivals access to data, although licenses would have to be designed to make the shared data useful without compromising privacy.

Compulsory licensing should be used sparingly to address industry dominance because it can diminish innovation incentives. An additional concern is the cost of administering compulsory licenses. The consent decrees that resolved the US and European Commission (EC) cases against Microsoft included obligations to license intellectual property and provide know-how necessary to achieve interoperability with Microsoft operating systems and applications. These are important forward-looking provisions, but they proved difficult to administer. The difficulties were related to narrow obligations for which Microsoft had discretion to specify licensing terms. Broader licenses with no or low royalties are easier to administer. While such obligations increase concerns about undermining innovation incentives, licensing obligation can be designed to promote follow-on innovation without allowing firms to imitate a market leader.

### The truncated rule of reason analysis is useful to assess allegations of anticompetitive innovation

One of the most challenging areas of antitrust enforcement for the high-technology economy is the evaluation of incremental innovations or product designs that allow a dominant firm to maintain a monopoly or extend it into related markets, but also have benefits for consumers. Courts and antitrust scholars have proposed various tests to analyze whether innovations that exclude competition should warrant antitrust scrutiny, but they have yet to agree on a preferred approach. For instance, the US FTC and the EC reached different conclusions in their evaluations of alleged bias in Google's displays of search results related to product comparison shopping services. The FTC did not challenge Google's conduct, which is consistent with a strong deference to innovation in US antitrust policy. The EC held that Google's conduct was unlawful, which is consistent with European policy that conduct by a dominant firm that discriminates against rivals can be a violation of European antitrust law. Neither approach strikes the right balance between promoting innovation incentives and protecting competition.

I suggest a truncated rule of reason to evaluate product designs and other innovations that may exclude competition. Under this approach, product designs and other innovations would escape antitrust condemnation if they offer substantial improvements and are not accompanied by other exclusionary conduct that does not have procompetitive benefits. Innovations and product designs that provide only marginal benefits would be examined under a full rule of reason analysis to determine whether their benefits compensate for any exclusionary effects.

This approach would not impose a large burden on potential innovators because the circumstances in which a marginal innovation or product design cause large exclusionary effects are limited. Some examples, such as pharmaceutical product hopping, have product changes and exclusionary effects that are no more difficult for courts to quantify than issues that courts address in many other antitrust cases. Exclusionary effects from changes in interoperability protocols are more complicated because the changes can have benefits and costs that are difficult to quantify. Nonetheless, the truncated rule of reason can accommodate these types of cases by imposing a low threshold for benefits from a change that defeats interoperability as a condition to require a balancing of costs and benefits.

### Antitrust authorities should evaluate the effectiveness of remedies for innovation

Antitrust authorities should not hesitate to impose harsh remedies, including structural divestitures, if they are warranted by the expectation of harm to innovation or future competition that cannot be satisfactorily addressed with more moderate measures. Unfortunately, there are no systematic studies of the performance of remedies that address innovation concerns. This book offers some anecdotes, but they paint a mixed picture. In the merger context, some divestitures appear to have successfully restored innovation incentives that allegedly would have been eliminated by the merger. Others appear to have been less successful: Recipients of the divested assets failed to invest in R&D directed toward applications for which there were innovation concerns. A successful remedy requires the identification of a party with the ability and incentive to restore investment allegedly lost from a challenged transaction and transfer of the R&D and related assets necessary to accomplish the desired ends.

Competition authorities have sponsored several studies of the effects of merger remedies on price competition in existing markets; however, little has been done to evaluate the extent to which remedies have succeeded in preserving innovation incentives and competition in markets for new products and services. Antitrust enforcement would benefit from comprehensive ex post reviews of the consequences that interventions have had for innovation and future competition. Indeed, antitrust enforcement would be served by a statutory obligation for merging firms to provide data on prices, R&D expenditures, patenting, and other relevant measures that would allow agencies to engage in retrospective analyses of the effects of mergers on competition and innovation.

### Should antitrust enforcement break up big tech?

I would be remiss if I did not at least briefly address the calls from many politicians and some antitrust scholars to break up major tech platforms. There are many concerns, in addition to perceived harm to innovation and price competition, that compel demands for structural reforms. They include the political power wielded by the giant platforms, the risks they pose to privacy, and monopsony power over workers and suppliers. The EC has levied substantial fines and ordered remedies to address perceived antitrust violations by the major tech platforms. However, these punishments and remedies have accomplished little to upset their market dominance.

I have no comparative advantage to assess whether structural reforms would alleviate concerns about political power or the misuse of personal data. However, it is worth noting that breaking up these firms does not necessarily imply that their successors would have greater interest in protecting consumer data or that regulators would have an easier time controlling abuses of privacy if more firms compete for the attention of consumers and advertisers.

Structural reform can be justified as an antitrust remedy to resolve persistent harm to competition that cannot be addressed satisfactorily with behavioral conditions, although it is easier to prevent combinations that create antitrust concerns in the first place. Structural reform often involves the separation of activities that have characteristics of a natural monopoly from other activities for which competition is feasible. That objective was an impetus for the separation of local telephone service from other potentially competitive services in the breakup of AT&T. The breakup promised that the local exchange companies, which would continue to be a regulated as monopolies, would not have incentives to foreclose or shift accounting costs onto services that could be provided under competitive conditions.

The major tech platforms are not regulated and consequently do not have the same anticompetitive incentives. Nonetheless, they have the ability and incentive to disadvantage firms that both compete with them and depend on access to their services. Google has the incentive and ability to disadvantage rivals that compete for search-based advertising revenues. Apple and Amazon have incentives to advantage their own products over other competitive products whose suppliers use their platforms to reach customers—and so on with respect to conduct by platforms in other dimensions.

Although the major tech platforms do not lack the ability and incentive to harm actual and potential rivals, the central question is whether structural divestiture is the best solution to address these concerns. Divestiture would require a determination of the boundary lines for the divested components and enforcement oversight to prevent the component parts from crossing into prohibited territories. Perhaps some acquisitions, such as Facebook's purchase of Instagram, could be unwound without major disruptions. But it is questionable whether unscrambling these eggs would have procompetitive consequences. It would not be difficult for Facebook to compete with an independent Instagram, and the inexorable forces of network effects are likely to produce a single victor. We can bet who that might be.

Other types of structural reforms might have more beneficial effects. For example, separating Amazon merchant services from sales of its privately branded products might alleviate concerns that Amazon would misuse data on merchant sales to favor its own products. But such reforms could sacrifice consumer benefits by prohibiting Amazon from supplying private label products at attractive prices, and by eliminating the use of Amazon's private label products to discipline the prices of related products. As a policy matter, it would also be difficult to explain why Amazon should be restricted from private label sales while allowing other sellers (such as Walmart) to make extensive use of this retail strategy. More complicated structural reforms, such as limiting the types of services that Google, Facebook, Amazon, or Apple can offer, could have benefits in theory, but they would be very difficult to design and enforce in ways that promote innovation and consumer welfare.

The design and administration of the divestiture of a major tech platform would be no easier than the breakup of AT&T. The AT&T divestiture was costly to administer and, more importantly, interfered with innovation that required coordination between the different parts of the divested Bell System. There is no reason to believe that the risk of harm to innovation would be less severe from a structural divestiture of any of the major tech platforms. Moreover, if a divestiture cleaves activities for which there are powerful network effects, the forces that enable a tech platform to sustain its dominance would tend to recreate a new dominant firm after a breakup.

Although there may be scope for informed structural change to address dominance in the digital economy, it is not a substitute for diligent antitrust enforcement. Antitrust enforcement for the high-technology economy can be strengthened by reforms proposed in this book, along with increased penalties for unlawful conduct and additional agency resources to develop the specialized expertise necessary to address harm to competition and innovation in this sector. Stronger antitrust enforcement should be joined with regulations designed to protect privacy and limit the political influence of these and other major companies. The purpose of this book is to provide a reference source of knowledge and experience to guide this diligent antitrust enforcement. The lessons are not simple; more needs to be done to understand the proper role of antitrust enforcement in promoting innovation for the high-technology economy.

# Acknowledgments

My interest in innovation began on the ground floor—literally the ground floor of a laboratory where I once did research on solid-state microwave devices. I was fascinated by the science, but over time I became even more fascinated by the problem of how we (meaning a research lab or a society) decide how to choose the science that merits our scarce investment resources. That was the start of my career in economics.

I was extraordinarily fortunate to gain the mentorship of Joseph Stiglitz and James Sweeney in my graduate education at Stanford. Joe had National Science Foundation support for a project on the economics of research and development (R&D). His project brought David Newbery to Stanford, where we collaborated on work that led to our paper on preemptive patenting and the persistence of monopoly.

An inflection point in my career came when Anne Bingaman, newly appointed by President Bill Clinton to be the Assistant Attorney General for Antitrust in the US Department of Justice (DOJ), asked me to be her economics deputy. Working in the Antitrust Division was an extraordinary experience, but I was surprised to see how little enforcement activity focused on innovation. Anne allowed me to direct a project that led to the joint publication by the DOJ and Federal Trade Commission (FTC) in 1995 of the Antitrust Guidelines for the Licensing of Intellectual Property. The guidelines were generally well received by antitrust scholars and practitioners and survived largely intact when the DOJ and the FTC issued a revised version in 2017.

This book is my attempt to provide a reference for scholarship on the topic of competition and innovation, along with relevant antitrust developments. It is informed by my experience over many years with scholars and practitioners in this field. They include my current and former Berkeley colleagues Alan Auerbach, Aaron Edlin, Joe Farrell, Steve Goldman, Ben Handel, Bronwyn Hall, Michael Katz, Jon Kolstad,

Rob Merges, Dan McFadden, Peter Menell, Aviv Nevo, Dan Rubinfeld, Suzanne Scotchmer, Carl Shapiro, Glenn Woroch, Hal Varian, Miguel Villas-Boas, Oliver Williamson, and Brian Wright. I am also fortunate to have worked in a teaching capacity at Berkeley with exceptional scholars who have made important contributions to competition policy and related fields and taught me as much as I might have taught them. They include Sofia Berto Villas-Boas, Leonard Cheng, Jeff Church, Alan Cox, Nick Economides, Eric Emch, Neil Gandal, Nancy Gallini, Justine Hastings, John Henly, Rene Kamita, Jacques Lawarree, Carmen Matutes, Nathan Miller, Yesim Orhun, Michael Riordan, Pierre Regibeau, Katherine Rockett, Andy Schwarz, Tim Simcoe, Ralph Winter, and Xavier Vives.

I am grateful to other colleagues in the economics and legal professions from whom I have learned much, including Reiki Aoki, Jon Baker, Jorge Contreras, Giulio Federico, Drew Fudenberg, Hillary Greene, Richard Harris, Ken Heyer, Andy Joskow, Paul Klemperer, Eirik Kristiansen, Tracy Lewis, Doug Melamed, Sam Miller, Aviv Nevo, Ariel Pakes, Christian Riis, Erlend Riis, Neil Roberts, Steve Sunshine, Jean Tirole, Will Tom, Greg Werden, and Michael Whinston. In my consulting work, I have benefited from interactions with Dennis Carlton, Bret Dickey, Meg Guerin-Calvert, Taylor Hines, Kun Huang, Gilad Levin, Janusz Ordover, Jon Orszag, Allan Shampine, Elizabeth Wang, Julie Wang, and Bobby Willig. Charles Clarke, Sanchit Shorewala, and Gwin Zhou have given me excellent research assistance.

Jon Baker, Sam Miller, Jim Ratliff, Greg Sivinski, and Glenn Woroch provided helpful comments on chapter drafts, and I absolve them of any errors. I owe a particular debt to Jim Ratliff, who has worked with me on many of the antitrust cases that I reference in this book and has been a continuing source of insight and inspiration.

And I thank Ray Riegert for his advice about book publishing, Alison Gilbert for suggesting the title of this book, and my family—and particularly Sandra—for their patience and support during this project.

# Notes

## Chapter 1

1. See, e.g., Khan (2017), Stiglitz (2017), and Wu (2018a, b).

2. I will refer to several actual antitrust cases to illustrate the principles in this book. I was personally involved in some of these cases, either in my capacity as Deputy Assistant Attorney General for Economics in the Antitrust Division of the DOJ or as a private consultant. In the interest of transparency, I indicate my involvement when I discuss the cases. My statements about these cases are not intended to reflect or endorse the official positions of any enforcement authority or any other party and do not reference confidential information. I have not received financial support from any private entity for the writing of this book.

3. I use the term "merger" here and throughout this book to also refer to acquisitions in which a firm acquires the assets of another firm or subsidiary and retains its preacquisition identity.

4. See, e.g., OECD (2018) at 20.

5. Ibid. at 44.

6. *US v. General Motors and ZF Friedrichshafen, AG, et al.*, Civil Action 93–530, Complaint, November 16, 1993.

7. The author was Deputy Assistant Attorney General for Economics when the Antitrust Division of the DOJ began its investigation of Microsoft and subsequently consulted with the Division in a private capacity.

8. 15 US Code § 1.

9. 15 US Code § 2.

10. 15 U.S.C. §§ 12-27, as amended.

11. 15 U.S.C. sec. 45(a)(1).

12. See, e.g., Hovenkamp (2010).

13. Canada introduced "An Act for the Prevention and Suppression of Combinations Formed in Restraint of Trade" in 1889, one year before the US passed the Sherman Act.

14. Arrow (1962).

15. Schumpeter (1942).

## Chapter 2

1. Isaacson (2014) provides a personalized account of the human factors that drove innovation for the digital economy.

2. This distinguishes applied innovation from basic research, for which there is a continued role for support from government, universities, and institutions that does not directly depend on commercial profitability.

3. See, e.g., Sidak and Teece (2009).

4. R&D intensity is likely to increase for motor vehicles that incorporate self-driving features, but perhaps not for automobile manufacturers if they purchase the technology from external suppliers.

5. Schumpeter (1942) at 83.

6. This is based on the rate at which firms enter and exit the S&P 500 Index. See Fox (2017).

7. In general, this presumption is correct (for existing resources) only if prices for all other goods and services equal their marginal costs. The theory of the second best says that marginal cost pricing is not necessarily efficient if prices for some products or services in the economy diverge from their marginal production costs. See Lipsey and Lancaster (1956).

8. Network effects are sometimes called network "externalities" to emphasize that consumers and firms may not account for the effects of their purchase and product decisions on the welfare of other consumers and firms.

9. See Katz and Shapiro (1985, 1994).

10. See Katz and Shapiro (1994) and Farrell and Saloner (1985, 1986).

11. Ibid.

12. Church and Gandal (1993) show that in markets for competing but incompatible computer systems, indirect network effects from complementary computer software can lead to inefficient adoption of computer hardware. Katz and Shapiro (1986a, b) and Farrell and Katz (1998, 2005) explore how consumer expectations can affect market competition with network effects.

13. See, e.g., Shelanski and Sidak (2001) at 35–36.

14. See, e.g., Lee (2014) and Weyl and White (2014).

15. Farrell and Katz (2005) at 237 conclude that "it seems almost inevitable that predation policy in network markets will sometimes be fighting the wrong war."

16. *US v. Microsoft*, Court of Appeals for the District of Columbia Circuit (June 28, 2001), 253 F.3d 34, 11–12.

17. The fifth firm is Berkshire Hathaway.

18. "The key feature of payment systems, and one that arises in several other industries characterized by network externalities (media, software, matchmakers, etc.), is its two-sidedness." Rochet and Tirole (2002); see also Baxter (1984–1985).

19. Rochet and Tirole (2006).

20. Evans and Schmalensee (2018) maintain that it is appropriate to view a business as a platform if (1) there are indirect network effects between members of at least one of the two customer groups and members of the other group, (2) these indirect network effects are strong enough to affect business conduct, and (3) the platform facilitates interactions between members of the two groups.

21. Lee (2013).

22. *Ohio et al. v. American Express Co. et al.*, US Supreme Court, No. 16-1454 (2018). (Note: The author testified on behalf of American Express in this case.)

23. The Supreme Court's majority opinion elicited a strong dissent, written by Justice Stephen Breyer, who argued that the district court was correct to focus its analysis on a market for credit card network services provided to merchants and that there was no precedent for defining a two-sided market. Ibid. at 14, 19.

24. Generally, platform competition is more intense on a side where agents are likely to choose a single platform (agents "single-home") rather than participate on several platforms (agents "multihome"). See, e.g., Armstrong (2006). Merchants that accept credit cards have an incentive to multihome because they want to serve any customer that presents a card. Many cardholders are willing to concentrate their purchases on a single card that offers high rewards if merchants accept the card, which incentivizes card issuers to offer high rewards.

25. See, e.g., Filistrucchi, Geradin, van Damme, and Affeldt (2014).

26. See, e.g., Cukier (2010).

27. See, e.g., Rubinfeld and Gal (2017).

28. See, e.g., Sivinski, Okuliar, and Kjolbye (2017).

29. US Department of Justice Press Release, "Justice Department requires Thomson to sell financial data and related assets in order to acquire Reuters," February 19, 2008.

30. *US v. Thomson Corporation and Reuters Group plc*, Complaint, February 19, 2008.

31. US Federal Trade Commission, *In the Matter of Nielsen Holdings N.V. and Arbitron Inc.*, File No. 131 0058, "Analysis of Agreement Containing Consent Order to Aid Public Comment," September 20, 2013.

32. European Commission, *Microsoft/LinkedIn*, Case M.8124, December 6, 2016 at §5.

33. Such firms are sometimes called "complementors." Shaprio and Varian (1999b) at 10.

34. Cournot (1927).

35. Integration often increases innovation effort compared to efforts by separate firms when the efforts are complementary. However, unlike the Cournot complements effect for prices, this is not a general result because there is an interaction with prices, which also affect the demand for innovations. See the appendix to Farrell and Katz (2000).

36. See US Department of Justice Press Release, "Justice Department will not challenge Cisco's acquisition of Tandberg," March 29, 2010; available at https://www.justice.gov /opa/pr/justice-department-will-not-challenge-cisco-s-acquisition-tandberg. (Note: The author consulted for Cisco Systems on matters related to this transaction.)

37. Cisco news release, "Cisco receives antitrust approvals from European Commission and US Department of Justice for pending acquisition of TANDBERG"; available at https://newsroom.cisco.com/press-release-content?type=webcontent&articleId=5430146, accessed March 12, 2019.

38. US Federal Trade Commission, *In the Matter of Intel Corporation*, Docket No. 9341, Complaint, December 16, 2009.

39. US Federal Trade Commission, *In the Matter of Intel Corporation*, Docket No. 9341, Decision and Order, October 29, 2010.

40. Not all firms in the high-technology economy face low barriers to enter related markets. A drug company that specializes in cardiovascular diseases would incur large expenses for R&D, clinical testing, and distribution to enter markets for vaccines and would have limited economies of scope from its existing assets.

41. Dryden and Iyer (2017) and Srinivason (2019) offer some other examples in which privacy issues intersect with antitrust concerns.

42. See, e.g., Furman and Orszag (2015), Furman (2016), and Gutiérrez and Philippon (2018).

43. See, e.g., Autor, Dorn, Hanson, Pisano, and Shu (2017) and Stiglitz (2017).

44. See, e.g., Khan (2017), Stiglitz (2017), and Wu (2018a, b).

45. See, e.g., Herndon (2019).

46. See, e.g., Baker and Shapiro (2008) and Kwoka, Greenfield, and Gu (2015).

47. Jay R. Ritter, Initial public offerings: Updated statistics; available at https://site .warrington.ufl.edu/ritter/files/2019/01/IPOs2018Statistics_Dec.pdf, accessed November 5, 2019.

48. Mauboussin, Callahan, and Majd (2017).

49. "Into the danger zone: American tech giants are making life tough for startups," *The Economist*, June 2, 2018, available at https://www.economist.com/business/2018/06/02 /american-tech-giants-are-making-life-tough-for-startups.

50. Del Rey (2017).

51. For example, Facebook acquired Onavo in 2013. Onavo harvests user data and provides analytics for mobile applications, which Facebook can use to identify promising upstarts and potential competitors. Facebook reportedly shut down Onavo in 2019 in response to concerns about its collection of user data. See, e.g., Wells (2019).

52. Crunchbase, "Number of Google acquisitions," available at https://www.crunchbase .com/organization/google/acquisitions/acquisitions_list#section-acquisitions, accessed September 16, 2019.

53. Google sold Motorola to Lenova a few years after the acquisition but kept most of the patents.

54. Geradin and Katsifis (2018).

55. The Verge, "Facebook is shutting down a teen app it bought eight months ago," July 2, 2018; available at https://www.theverge.com/2018/7/2/17528896/facebook-tbh -moves-hello-shut-down-low-usage, accessed November 15, 2019.

56. Stone (2013) details a fierce price war between Amazon and Quidsi that ended when Amazon purchased Quidsi for $545 million in 2010. Amazon shut down the Quidsi websites several years after the acquisition, although Amazon continues to sell consumables on its merchant platform.

57. "Hyenas and cheetahs: Artificial intelligence has revived the semiconductor industry's animal spirits," *The Economist*, June 9–15, 2018, pp. 54–56.

58. European Commission, *Facebook/WhatsApp*, Case No. COMP/M.7217, October 3, 2014.

59. UK Office of Fair Trading, "Anticipated acquisition by Facebook Inc of Instagram Inc," ME/5525/12, 14 August 2012.

60. Market Realistic, available at https://articles2.marketrealist.com/2019/01/instagrams-ad-revenue-more-than-doubled-in-2018/, accessed November 5, 2019.

61. US Federal Trade Commission, "Statement of the Commission concerning Google/AdMob," FTC file no. 101–0031, May 21, 2010. See also Farrell, Pappalardo, and Shelanski (2010) at 266.

62. eMarketer, "Net US mobile ad revenue share, by company, 2014–2017"; available at https://www.emarketer.com/Chart/Net-US-Mobile-Ad-Revenue-Share-by-Company-2014-2017-of-total-billions/176289, accessed January 13, 2019.

63. Mobbo, home page; available at https://mobbo.com/whitepaper-monetization/, accessed February 19, 2019.

64. Cunningham, Ederer, and Ma (2018).

65. Hovenkamp (2013) at 2471: "The dominant view of antitrust policy in the United States is that it should promote some version of economic welfare. More specifically, antitrust promotes allocative efficiency by ensuring that markets are as competitive as they can practicably be and that firms do not face unreasonable roadblocks to attaining productive efficiency, which refers to both cost minimization and innovation."

66. The Antitrust Modernization Commission was created pursuant to the Antitrust Modernization Commission Act of 2002, Pub. L. No. 107–273, 116 Stat. 1856.

67. Garza et al. (2007) at 9. See also Posner (2001) at 925: "Antitrust doctrine is supple enough, and its commitment to economic rationality strong enough, to take in stride the competitive issues presented by the new economy."

68. 15 US Code § 18.

69. 353 US 586, 592.

70. 463 F. Supp. 983 (December 29, 1978). See Tom (2001) for a useful discussion of the case.

71. 463 F. Supp. 983, 1000.

72. 645 F.2d 1195, 1210 (March 12, 1981) (emphasis added).

73. *Golden Gate Pharmacy Services, Inc. v. Pfizer, Inc.*, No. C-09–3854 (N.D. Cal. Dec. 2, 2009).

74. *Golden Gate Pharmacy Services, Inc. v. Pfizer, Inc.*, Court of Appeals for the Ninth Circuit, 433 Fed. Appx. 598 (May 19, 2011).

75. See Royall and DiVincenzo (2010). The doctrine of "actual potential competition" differs from the doctrine of "perceived potential competition." The latter refers to the

disciplining effect of a firm that could enter an industry on prices charged by incumbents. The law has been more receptive to arguments about perceived potential competition than actual potential competition.

76. See, e.g., Kolasky and Dick (2003) and Kwoka (2008).

77. Hovenkamp (2013) at 2476.

## Chapter 3

1. Joseph Schumpeter does not present a formal description of his theory about the relationship between competition and innovation. His views appear in several publications but are most prominent in *Capitalism, Socialism, and Democracy* (1942).

2. Arrow (1962).

3. Both Kenneth Arrow and John Hicks were awarded the Nobel Prize in Economics in 1972. The Nobel committee emphasized the two men's pioneering contributions to general equilibrium and welfare theory rather than competition and innovation, but Arrow's Nobel biography highlights his 1962 paper on the allocation of resources to invention, and it is one of his most-cited papers. The Nobel Prize, "Kenneth J. Arrow, Biographical," available at https://www.nobelprize.org/prizes/economic-sciences /1972/arrow/biographical/.

4. Schumpeter (1942) at 106. Since impossible implies inferior, a more logical interpretation of this well-known quote is that competition would be an inferior organizational form for economic progress if it were possible.

5. Schumpeter (1942) at chapter VII. McCraw (2012) disputes the notion that Schumpeter promoted the benefits of size and monopoly power for innovation.

6. Schumpeter (1942) at 100–101 (footnote omitted).

7. Dasgupta and Stiglitz (1980).

8. Vives (2008).

9. See Spence (1984) for an analysis of cost-reducing innovation incentives with technological spillovers. Spillovers partly explain why geographic clusters of firms tend to be associated with high levels of industry productivity. See Porter (2001).

10. Katz and Shapiro (1987).

11. Arrow (1962) at 620.

12. The social value includes spillover benefits for firms and consumers that innovators often do not capture as profits. Jones and Williams (1998) estimate social rates of return to innovation and find that they average in excess of 27 percent per annum, which is much more than the average private return on innovation.

13. The discussion of monopoly preemption incentives in this chapter is based on Gilbert and Newbery (1982).

14. Isaacson (2011) at 408.

15. Reinganum (1983, 1989) makes this point clearly in a dynamic model of a race to patent an invention.

16. See Salant (1984).

17. Lewis (1983) shows that the value of a preemption strategy is a decreasing function of the number of alternative entry paths, and preemption is not profitable if the number of paths is sufficiently large.

18. The *US Department of Justice and Federal Trade Commission Antitrust Guidelines for the Licensing of Intellectual Property* do not presume that a patent, copyright, or trade secret necessarily creates market power for its owner, let alone confers a monopoly. The guidelines cite *Ill. Tool Works Inc. v. Indep. Ink, Inc.*, 547 US 28, 45–46 (2006) ("Congress, the antitrust enforcement agencies, and most economists have all reached the conclusion that a patent does not necessarily confer market power upon the patentee. Today, we reach the same conclusion.")

19. Shapiro and Varian (1999) offer examples of industries in which complements feature prominently and describe implications for profitable business strategies.

20. Greenstein and Ramey (1998).

21. Chen and Schwartz (2013).

22. See, e.g., *Economic Report of the President*, February 2016.

23. Coase (1972) observed that the supplier of a durable good has an incentive to lower its prices continually to increase its sales. The distinction between pricing incentives for a durable and a nondurable good would be lost if the supplier of the durable good rented rather than sold it.

24. Ellison and Fudenberg (2000) show that a durable good monopolist can have excessive incentive to supply upgrades.

## Chapter 4

1. Winner-take-all is a strong assumption in this context. It rules out sharing or the expectation of sharing from simultaneous discovery.

2. This result first appears in Sah and Stiglitz (1987).

3. Total R&D investment can exceed the level that maximizes total economic welfare (the welfare of consumers and the profits earned by firms) because the incentive to innovate to take business from a rival does not have a corresponding social benefit.

4. Levin, Klevorick, Nelson, and Winter (1987) and Cohen, Nelson, and Walsh (2004) describe the mechanisms that firms employ to protect their comparative advantage.

5. See Loury (1979), Lee and Wilde (1980), Reinganum (1981), and the survey in Reinganum (1989). Technically, these models assume a Poisson discovery probability with a hazard rate that is independent of past R&D expenditures.

6. See, e.g., Jones and Williams (1998).

7. Stewart (1983) develops a model of a patent race and finds that intermediate levels of concentration can promote innovation if rivals benefit from technological spillovers.

8. Vickers (1986).

9. Fudenberg, Gilbert, Stiglitz, and Tirole (1983) and Harris and Vickers (1985) show that monopoly outcomes in R&D can emerge under more general conditions when firms differ in their stock of R&D capital and they can monitor investments by their rivals.

10. Doraszelski (2003) generalizes the patent race model to allow discovery probabilities that depend on past R&D expenditures. His model allows profit-maximizing strategic behavior, in which a firm that lags a technological leader can catch up to the leader by aggressively increasing its expenditures on R&D, provided that it already has a sufficiently large stock of R&D capital. The model also allows for increasing dominance, in which a technological leader builds on its lead and the laggard falls further behind.

11. Variants of the stepwise model include Aghion, Harris, and Vickers (1997), Aghion, Harris, Howitt, and Vickers (2001), and Aghion, Bloom, Blundell, Griffith, and Howitt (2005).

12. One example of this was done by varying the elasticity of substitution for the duopolists' products.

13. Biotech Industry Report (2015).

14. Erin Griffith, "Will Facebook kill all future Facebooks?," *Wired*, available at https://www.wired.com/story/facebooks-aggressive-moves-on-startups-threaten-innovation/, accessed October 18, 2019, and the discussion in chapter 2.

15. For example, suppose the innovation is a nondrastic reduction in marginal cost and the price elasticity of substitution is equal and constant for all products in the acquiring market. Then an increase in the number of potential buyers with the old technology lowers the value of acquiring the new technology.

16. Marshall and Parra (2019).

17. The quoted phrase is in a letter from Isaac Newton to Robert Hooke, reported in Koyré (1952) at 5.

18. See, e.g., Merges and Nelson (1990).

19. Merges and Nelson (1990). The pool created the Radio Corporation of America (RCA). Patent logjams occurred in the early development of other industries, including electric lighting and automotive technology.

20. Research papers that examine the implications of cumulative innovation for the design of intellectual property rights include Scotchmer (1991, 2004), Chang (1995), O'Donoghue (1998), O'Donoghue, Scotchmer, and Thisse (1998), and Hunt (2004).

21. Kitch (1977).

22. "To give the second innovator an incentive to invest whenever social benefits exceed R&D costs, the second innovator must earn the entire social surplus of his innovation. But to compensate the first innovator for the externality or spillover she provides, she too must earn part of this surplus. It is impossible to give the surplus to both parties" (Scotchmer, 1991 at 34). See also Scotchmer (2004), at chapter 5.

23. The acronym stands for "clustered regularly interspaced short palindromic repeats." A palindromic sequence is a strand of DNA that is a mirror image of a sequence of nucleic acids in the complementary strand.

24. See Gilbert and Kristiansen (2018).

25. A further complication is that new competitors can become established firms whose conduct affects rewards for subsequent innovators. See, e.g., Segal and Whinston (2007) and Baker (2016).

26. Bessen and Maskin (2009).

27. See, e.g., Hart (1983).

28. See, e.g., Holmstrom (1982) and Nalebuff and Stiglitz (1983).

29. Scharfstein (1988) showed that product competition can lead to optimal compensation schemes that increase or decrease managerial effort depending on managers' preferences. Hermalin (1992) and Schmidt (1997) showed that competition can harm the ability of owners to negotiate compensation schemes with executives that motivate profit-maximizing behavior.

30. Dasgupta and Maskin (1987).

31. Cabral (1994) and Kwon (2010).

32. Christensen (1997).

33. Joshua Gans defines a technology as disruptive if the choices that once drove a firm's success become those that destroy its future. He also argues that disruptive technologies are inherently unpredictable, or else incumbents would react and avoid destructive consequences. Gans (2016) at 9.

34. See, e.g., Lepore (2014) and King (2017). These authors suggest that Christensen does not adequately account for mergers and acquisitions that enabled firms to sustain industry dominance.

35. This discussion is based on Tripsas and Gavetti (2000).

36. See Gilbert and Newbery (1982) and the discussion in chapter 3.

37. Franco and Filson (2006).

38. See, e.g., Gans (2016) at 122, 126.

39. Henderson and Clark (1990).

40. See, e.g., Lynn (1998).

41. While innovations sometimes require costly and disruptive changes to incumbents' existing competencies, incumbents can have differential advantages for other innovations because they complement their existing competencies. See, e.g., Tushman and Anderson (1986).

42. Grove (1996) at 106. Grove explained that the market success of 486-based computers was a crucial factor in Intel's decision to focus its efforts on the existing CISC microprocessor architecture.

43. Henderson (1993).

44. See, e.g., "Apple doesn't rely on market research, says marketing chief Phil Schiller," *Appleinsider*, July 31, 2012; available at https://appleinsider.com/articles/12/07/31/apple_doesnt_rely_on_market_research_says_marketing_chief_phil_schiller, accessed December 3, 2019.

45. Gans (2016) describes strategies that incumbent firms have employed to manage disruptive change.

46. Hawking (2005) at 121, 132.

## Chapter 5

1. I do not distinguish between mergers and asset acquisitions in this discussion.

2. Some treatises classify acquisitions of potential competitors as a vertical transaction, but I do not follow that convention in this book.

3. Vertical mergers can have efficiency benefits that are generally absent in horizontal mergers, such as reducing mark-ups in the supply chain. See, e.g., European Commission (2016) at 4.

4. William Baxter, the Assistant Attorney General for Antitrust in the administration of Ronald Reagan, noted that mergers can raise prices for "today's products," for "tomorrow's products," or for the R&D activity that creates new or improved products (Baxter, 1984–1985).

5. See the Hart-Scott-Rodino Annual Report Fiscal Year 2018. The Hart-Scott-Rodino Antitrust Improvements Act of 1976, Pub. L. No. 94–435 ("HSR Act"), obligates companies to report proposed mergers or acquisitions to the FTC if they exceed defined monetary thresholds. Proposed mergers or acquisitions that fall below the HSR thresholds are not exempt from antitrust review, although reporting greatly facilitates review by the antitrust agencies.

6. The DOJ and FTC *Antitrust Guidelines for the Licensing of Intellectual Property* ("IP Guidelines") published in 1995 and revised in 2017 did not address merger policy, but they noted the possible impacts of licensing arrangements on innovation incentives, and their publication in 1995 coincided with an inflection point in merger enforcement for innovation.

7. See Hesse (2014). In a footnote, she acknowledges that price and innovation effects do not always go hand in hand.

8. See Gilbert and Tom (2001), who observe that innovation concerns were not pivotal to enforcement decisions that occurred prior to the turn of the twentieth century.

9. The 1968 Merger Guidelines said, "In certain exceptional circumstances, however, the structural factors used in these guidelines [to define markets] will not alone be conclusive.... This is sometimes the case, for example, where basic technological changes are creating new industries, or are significantly transforming older industries, in such fashion as to make current market boundaries and market structure of uncertain significance."

10. Merger Guidelines (1982), reprinted in 4 Trade Reg. Rep. ¶ 13,102 § I (June 14, 1982), available at www.justice.gov/atr/hmerger/11248.htm.

11. A 1992 revision of the Merger Guidelines went slightly further to note that non-price impacts may include "product quality, service, or innovation."

12. See US Department of Justice and Federal Trade Commission (2010) at § 10 ("Cognizable efficiencies are merger-specific efficiencies that have been verified and do not arise from anticompetitive reductions in output or service").

13. Oliver Williamson (1983) describes advantages of internal organization compared to market transactions. In addition to his distinguished academic career, Williamson was a chief economist in the Antitrust Division of the DOJ. He was awarded a Nobel Prize in 2009.

14. Denicolò and Polo (2018a, 2018b) show that, under some conditions, merging parties can profitably reposition their R&D assets such that their innovation incentives post-merger exceed the parties' innovation incentives prior to the merger.

15. US Department of Justice and Federal Trade Commission (2010) at § 10.

16. US Department of Justice, Statement of the Department of Justice Antitrust Division on Its Decision to Close Its Investigation of the Internet Search and Paid Search Advertising Agreement Between Microsoft Corporation and Yahoo! Inc. (February 18, 2010).

17. US Department of Justice, Department of Justice Antitrust Division Statement on the Closing of Its Investigation of the T-Mobile/MetroPCS Merger (March 12, 2013).

18. See European Commission (2004) at ¶ 8.

19. Ibid. at ¶ 38.

20. For example, Crane (2011) argues for a more symmetrical treatment of predicted harms and benefits in merger enforcement.

21. Baker and Shapiro (2008) allege that some courts and enforcers have allowed mergers to proceed based upon dubious economic arguments about concentration, entry, expansion, and efficiencies.

22. See the review of merger retrospectives in Kwoka, Greenfield, and Gu (2015). Based on these studies, the authors argue that the agencies have not been sufficiently aggressive in challenging mergers and that structural and conduct remedies have not prevented anticompetitive price increases.

23. Michael Katz and Howard Shelanski call the causal effect of market structure on innovation the "innovation incentives" effect and call the reverse effect of innovation on market structure the "innovation impact" effect. See Katz and Shelanski (2005, 2007b).

24. Pleatsikis and Teece (2001) argue that the conventional approach to merger analysis ignores the dynamic nature of competition, which constrains market power for high-tech firms.

25. See, e.g., Posner (2001) and the discussion in chapter 2 of this book.

26. The DOJ did not challenge the merger of the satellite radio companies XM and Sirius, in part because the Antitrust Division anticipated competition from new audio-streaming services (US Department of Justice, Statement of the Department of Justice Antitrust Division on its Decision to Close its Investigation of XM Satellite Radio Holdings Inc.'s Merger with Sirius Satellite Radio Inc., March 24, 2008). (Note: The author consulted with the Antitrust Division on the XM-Sirius merger.)

27. See, e.g., the discussion of *US v. Microsoft* in chapters 2 and 8 (dynamic markets do not preclude investigations of anticompetitive conduct).

28. I use the terms "tacit agreement" and "conscious parallelism" interchangeably. Some characterize tacit agreement as an actual agreement that is not codified as an oral or written expression. This contrasts with conscious parallelism, in which firms recognize the interdependence of their actions, such as restraining from cutting prices because rivals will cut their prices.

29. One survey of R&D lab managers and directors found that about 85 percent of respondents did not learn about rivals' R&D projects until either the development stage or subsequent to product introduction (Cohen, Nelson, and Walsh, 2004).

30. *US v. Automobile Manufacturers Ass'n*, 307 F. Supp. 617 (C.D. Cal. 1969), aff'd in part and appeal dismissed in part; *City of New York v. US*, 397 U.S. 248 (1970).

31. European Commission Press Release, "Antitrust: Commission fines truck producers € 2.93 billion for participating in a cartel," Brussels, July 19, 2016.

32. *Kaufman v. BMW*, 17-cv-05440, U.S. District Court of New Jersey (July 25, 2017), and *Burton v. BMW AG*, 17-cv-04314, U.S. District Court, Northern District of California (July 28, 2017).

33. See Farrell and Shapiro (2010).

34. Ibid. at 33.

35. See Bulow, Geanakoplos, and Klemperer (1985).

36. Ibid.

37. Examples of unilateral effects models of innovation incentives include papers by Letina (2016), Federico, Langus, and Valletti (2017, 2018), Motta and Tarantino (2017), López and Vives (2019), and Gilbert (2019).

38. See, e.g., Salinger (2016) and Gilbert (2019).

39. As noted in chapter 4, Sah and Stiglitz (1987) first demonstrated that innovation incentives are independent of market structure if the discovery is winner-take-all. See also Gilbert (2019) and Jullien and Lefouili (2018).

40. D'Aspremont and Jacquemin (1988) provide an example in which a merger to monopoly increases both R&D and total economic welfare (but not consumer welfare) if intrafirm spillovers are sufficiently large. See also Motta and Tarantino (2017).

41. See López and Vives (2019) for a thorough analysis of the effects of intrafirm spillovers and competition on firms' incentives to invest to lower their production costs.

42. See Aghion, Harris, Howitt, and Vickers (2001) and Aghion, Bloom, Blundell, Griffith, and Howitt (2005).

43. See, e.g., Federico, Scott-Morton, and Shapiro (2020).

44. See Gilbert (2019).

45. Royall and Divincenzo (2010) document the agencies' low success rates for these types of cases.

46. A district court denied the FTC's most recent attempt to prevent the acquisition of a potential competitor for this reason. *FTC v. Steris Corporation*, 133 F. Supp. 3d 962 (September 25, 2015).

47. DiMasi, Grabowski, and Hansen (2016) found that 1,442 self-originated compounds from the top fifty pharmaceutical firms had an average probability of clinical success of about 12 percent. Furthermore, a sample of 106 randomly selected new drugs developed by ten pharmaceutical firms required, on average, about eight years to complete clinical trials.

48. For example, sildenafil citrate began clinical trials for the treatment of hypertension and angina, but now it is better known as the erectile dysfunction drug Viagra.

49. The measure of expected harm $H$ is net of any merger-specific efficiencies that would be realized postmerger, under the assumption that the project would have been successful if Alpha and Beta had not merged. $H$ may include any decrease or increase in the discovery probability from the merger.

50. This approach to evaluate the effects of mergers that eliminate potential competitors is consistent with the decision-theoretic approach advocated by Katz and Shelanski (2007a). They observe that courts typically require evidence that the claimed harms and efficiencies exceed probability thresholds, which can lead to different enforcement outcomes than would be achieved from decision theory.

51. *United States v. Phila. Nat'l Bank*, 374 US 321, 361 (1963). ("Specifically, we think that a merger which produces a firm controlling an undue percentage share of the relevant market, and results in a significant increase in the concentration of firms in that market, is so inherently likely to lessen competition substantially that it must be enjoined in the absence of evidence clearly showing that the merger is not likely to have such anticompetitive effects.")

52. IP Guidelines (January 12, 2017) at 11–12 (footnote omitted).

53. See, e.g., Carlton and Gertner (2003) and the response in Hoerner (1995) to Gilbert and Sunshine (1995).

54. IP Guidelines at 25.

55. US Department of Justice and Federal Trade Commission (2000) at § 4.3.

56. Katz and Shelanski (2007b) promote a much broader safe harbor. They conclude that economic evidence is weak for the innovation effects of mergers, with the possible exception of merger to monopoly, although they note that mergers can have replacement and business-stealing effects that harm innovation incentives.

57. Shapiro (2012) calls this "contestability."

58. "If we see R&D assets only dimly, we see tomorrow's product market more dimly still.... All the usual difficulties of measuring market shares are compounded, as we attempt to extend our time horizon farther into the future" (Baxter, 1984).

59. See Cartwright and Ahmed (2016) at 4.

60. See the discussion of the proposed acquisition of HeartWare by Thoratec in chapter 7.

## Chapter 6

1. See, e.g., Cohen (2010), Damanpour (2010), Gilbert (2006), and Baldwin and Scott (1987).

2. See, e.g., Scherer (1965).

3. Hall and Ziedonis (2001).

4. Henderson (1993).

5. This circularity also arises in empirical studies that attempt to determine a relationship between concentration and profits. A firm can have a very large market share because it charges very low prices, but that does not mean that market concentration causes low prices. Rather, the firm's choice to offer low prices creates the high degree of market concentration.

6. See, e.g., Levin, Cohen, and Mowery (1985).

7. John Sutton (1998) developed the theory that costs and characteristics of market demand, particularly the ability of firms to use innovation to raise consumers' willingness to pay for their products, imply a lower bound on the level of industry concentration and the ratio of R&D investment to sales.

8. Cohen (2010).

9. Philippe Aghion and Jean Tirole call the relationship between R&D (or the output of innovations) and variables that alter the incentives for R&D "the second most tested hypothesis in industrial organization" after the relationship between market structure and prices (Aghion and Tirole, 1994 at 1195).

10. "Total factor productivity" is the portion of output that is not explained by labor and capital. It is measured by dividing output by the weighted average of labor and capital inputs.

11. See, e.g., Syverson (2011) and Van Reenen (2011).

12. Ghosh (2001) and Blonigen and Pierce (2016) find no evidence of merger efficiencies from operating data. Devos et al. (2016) report merger efficiencies based on forecasts by financial analysts.

13. See, e.g., Nickell (1966).

14. Blundell, Griffith, and Van Reenen (1999).

15. Damanpour (2010) surveys empirical studies and finds that most report a positive relationship between firm size and both product and process innovation.

16. See Aghion, Bloom, Blundell, Griffith, and Howitt (2005).

17. The Lerner Index is $L = (p-c)/p$, where $p$ is the firm's price and $c$ is its marginal cost. Studies typically use average variable cost as a surrogate for marginal cost.

18. Aghion, Blundell, Griffith, Howitt, and Prantl (2009).

19. Instead of citation-weighted patents, the researchers use patents by UK firms registered with the US Patent and Trademark Office, which they argue are more likely to be highly valued due to the cost of obtaining and maintaining foreign patents.

20. Bloom, Draca, and Van Reenen (2016).

21. Unsurprisingly, they find that low-wage competition from China correlated with the exit of many low-tech manufacturing firms and the reallocation of employment to higher-technology production methods. This is the Darwinian selection effect from competition on productivity. They also report an increase in total factor productivity within surviving firms, most of which were relatively advanced, as measured by total factor productivity.

22. Gutiérrez and Philippon (2017) also document a trend toward increasing market concentration and profitability in the US, which they attribute to relatively low investment by market leaders relative to their profits.

23. Autor, Dorn, Hanson, Pisano, and Shu (2020) at 16.

24. See, e.g., Hombert and Matray (2018) and Xu and Gong (2017). Relatedly Kueng, Li, and Yang (2016) find a larger negative impact from competition on process innovation, for which benefits are proportional to firm scale. Macher, Miller, and Osborne (2017) find that larger firms are more likely to adopt cost-saving innovations in the cement industry. Adoption is less likely when firms face many nearby competitors, which the authors attribute to depriving firms of the scale necessary to recoup adoption costs.

25. Reinganum (1984).

26. See Gilbert and Newbery (1982) and the discussion in chapter 3.

27. Christiansen (1997). Chapter 4 describes Christiansen's theory in more detail.

28. The HDD industry demonstrates "winner-take-all" or "winner-take-most" competition. Consequently, this industry has the feature of a patent race, notwithstanding that patents have been relatively unimportant in this industry because the most relevant innovations are improvements in manufacturing processes to increase the density of data storage on the drive and read/write speeds. Firms are reluctant to patent process technologies because infringement is difficult to detect and enforce. One exception is a broad patent on 3.5-inch HDDs that was awarded in 1986, but ultimately held to be invalid (Igami, 2017).

29. Iian Cockburn and Rebecca Henderson (1994) find no evidence of racing behavior in their study of R&D investment in the ethical drug industry. Instead, they conclude that firms' investment decisions were driven by heterogeneous capabilities, adjustment costs, and the evolution of technological opportunity, with a strong influence from technological spillovers.

30. By renting rather than selling a durable good, the seller can use price to influence the demand for rentals by customers who have previously rented the good. Thus, a rental market gives a durable good the characteristic of a nondurable good with respect to consumer demand.

31. Lohr and Kantor (2019).

32. Commission Press Release IP/09/745 of May 13, 2009, available at http://europa.eu/rapid/press-release_IP-09-745_en.htm?locale=en. The European Court of Justice subsequently remanded the case for further consideration. See Court of Justice of the European Union Press Release No. 90/17, September 6, 2017, available at https://curia.europa.eu/jcms/upload/docs/application/pdf/2017-09/cp170090en.pdf.

33. *In the Matter of Intel Corporation*, available at https://www.ftc.gov/enforcement/cases-proceedings/061-0247/intel-corporation-matter.

34. See Cabral (1994), Kwon (2010), and the discussion in chapter 4.

35. Henderson (1993).

36. Cohen (2010).

37. Garcia-Macia, Hsieh, and Klenow (2017).

38. See Gilbert and Greene (2015) and chapter 5.

39. Wollmann (2019) finds that the agencies fail to review many mergers that fall below the HSR reporting thresholds and that firms are more likely to pursue mergers that greatly increase market concentration when the transactions fall below the reporting thresholds.

40. Gutiérrez and Philippon (2017).

41. Cassiman, Colombo, Garrone, and Veugelers (2005).

42. Igami and Uetake (2018).

43. Sometimes a single patent covers a pharmaceutical product, while in many other industries (such as computers and information technologies), hundreds of patents cover a product.

44. DiMasi, Grabowski, and Hansen (2016).

45. For data on market concentration and R&D expenditures in the field of agricultural crop protection, see Fuglie et al. (2011) and Phillips McDougall (2016).

46. Haucap, Rasch, and Stiebale (2019).

47. The authors also find that acquisitions that fell below the reporting thresholds under the HSR Act were more likely to be terminated. This finding underscores the importance of merger reporting for effective antitrust enforcement and is consistent with the results reported by Wollmann (2019).

48. Schmalensee (1999) at 1326.

49. Antitrust Subcommittee (1959) at 38. Among other conditions, the decree also prohibited Western Electric, the manufacturing arm of the Bell System, "from manufacturing and selling equipment not of a type sold to the telephone operating companies of the Bell System" (Antitrust Subcommittee, 1959 at 37).

50. *United States v. International Business Machines*, Final Judgment, US District Court for the Southern District of New York, Civil Action No. 72–344 (1956).

51. *Xerox Corp.*, 86 F.T.C. 364. Licensees could designate up to three Xerox patents royalty-free, with a maximum royalty for all patents not to exceed 1.5 percent of the licensee's net revenues.

52. Willard Tom observed that "many of the practices alleged in the complaint or prohibited by the order seem innocuous to modern eyes and thus suggest an entirely foreign way of looking at the world" (2001 at 967).

53. Watzinger, Fackler, Nagler, and Schnitzer (2017).

54. Wessner (2001) at 86.

55. See Isaacson (2014) at 149.

56. Mowery (2011).

57. Grindley and Teece (1997) at 13.

58. See, e.g., Sabety (2005).

59. Tom (2001) at 989.

60. Bresnahan (1985).

61. Kearns and Nadler (1992).

62. Galasso and Schankerman (2015). Patent applicants are required to cite relevant patents even if they have been invalidated and the claimed technologies are in the public domain. In a subsequent study (Galasso and Schankerman, 2018), the authors find that invalidation of a patent on a core technology had a negative effect on future innovation by small and medium-size firms, but no significant impact for large firms.

63. Moser and Voena (2012).

64. Baten, Bianchi, and Moser (2017).

65. Scherer (1977).

66. Chien (2003). Compulsory licensing had no evident effect on innovation for five of six firms examined by Chien that were subjected to compulsory licensing decrees. One firm cut back on R&D investment, but she cites independent explanations for that behavior.

67. See, e.g., Delrahim (2004).

## Chapter 7

1. A nonexclusive license does not prevent the licensor from using the licensed technology or from licensing the technology to others.

2. See, e.g., the merger retrospective study by Kwoka, Greenfield, and Gu (2015) and the FTC merger divestiture study (US Federal Trade Commission, 2017).

3. See the related discussion of compulsory licensing in chapter 6.

4. *US v. General Motors and ZF Friedrichshafen*, Civil Action 93-530, Complaint, November 16, 1993 ("GM-ZF complaint") at ¶ 43.

5. The author consulted with the FTC in its evaluation of the proposed merger of Thoratec and HeartWare.

6. US Federal Trade Commission, *In the Matter of Thoratec Corporation and HeartWare International, Inc.*, Docket No. 9339, Complaint, July 28, 2009.

7. Hill, Rose, and Winston (2015).

8. Ibid. at 434.

9. Ibid. at 432.

10. Applied Materials 2017 Annual Report at 32. The "E" stands for engineering.

11. Tokyo Electron 2017 Annual Report.

12. See the discussion of Cournot complements in chapter 2.

13. Statement of Acting Assistant Attorney General Renata Hesse, US Department of Justice Press Release, "Lam Research Corp. and KLA-Tencor Corp. Abandon Merger Plans," October 5, 2016.

14. See comments by Dan Hutchenson of VSLI Research, quoted in Clark and Minaya (2016).

15. Firms that supply complementary products have a vertical relationship. The DOJ questioned whether behavioral commitments are adequate to address competition concerns in vertical mergers in its unsuccessful challenge in 2018 to the merger of AT&T and Time Warner (Delrahim, 2017).

16. US Federal Trade Commission Press Release, "FTC accepts proposed consent order in Broadcom Limited's $5.9 billion acquisition of Brocade Communications Systems, Inc.," July 3, 2017.

17. *US v. Lockheed Martin Corp. and Northrop Grumman Corp.*, Complaint, March 23, 1998.

18. US Federal Trade Commission, Statement of Chairman Robert Pitofsky and Commissioners Janet D. Steiger, Roscoe B. Starek III, and Christine A. Varney in the Matter of the Boeing Company/McDonnell Douglas Corporation, July 1, 1997.

19. US Federal Trade Commission Press Release, "FTC seeks to block Cytyc Corp.'s acquisition of Digene Corp.," June 24, 2002.

20. Department of Justice Press Release, "Justice Department reaches settlement with Microsemi Corp.," August 20, 2009.

21. CLP Buyer's Guide, Cytyc Corp., available at http://www.clpmag.com/buyers -guide/listing/cytyc-corp/, accessed July 23, 2018.

22. CLP Buyer's Guide, Digene Corp., accessed July 23, 2018.

23. See https://www.microsemi.com/products/ and http://www.semicoa.com/pro ducts/, accessed July 23, 2018.

24. US Federal Trade Commission, Statement of Chairman Timothy J. Muris in the matter of Genzyme Corporation/Novazyme Pharmaceuticals, Inc., January 13, 2004.

25. John Crowley's struggle to find a cure for Pompe disease is told in Anand (2006) and in the movie *Extraordinary Measures*.

26. Cunningham, Ederer, and Song (2018). This statistic does not demonstrate a loss of innovation, though, because the inventors may have made important contributions to innovation at other entities.

27. The reader might question why the FTC appears more often in these discussions despite the fact that the FTC and the Antitrust Division of the DOJ share responsibilities for merger enforcement. The reason is that cases with clearly identifiable innovation issues often appear in the pharmaceutical and related health industries, for which the FTC has historically been the lead agency for merger enforcement.

28. US Federal Trade Commission, "Announced Actions for October 4, 1996," available at https://www.ftc.gov/news-events/press-releases/1996/10/announced-actions-october-4 -1996, accessed November 3, 2019.

29. GC Pharma, "R&D Pipeline," available at http://www.globalgreencross.com/rd /pipeline, accessed April 9, 2018.

30. US Federal Trade Commission, *In the Matter of Glaxo Wellcome and SmithKline Beecham*, Analysis of Proposed Consent Order to Aid Public Comment (2000).

31. Ibid.

32. Gilead Sciences, Inc., Form 10-K, for the fiscal year ending December 31, 2001.

33. OSI Pharmaceuticals, Inc., Form 10-K, for the fiscal year ending September 30, 2004.

34. See Aurelian (2004).

35. Scrip, Pharma Intelligence, "Xenova acquires Cantab in stock swap," March 2001, available at https://scrip.pharmaintelligence.informa.com/deals/200110027, accessed October 5, 2019.

36. US Securities and Exchange Commission, "Celtic Pharma Development UK plc, Form T-3," September 12, 2005.

37. GSK Press Release, "GSK provides update on Herpevac trial for women evaluating Simplirix™ (Herpes Simplex Vaccine)," available at https://www.gsk.com/en-gb /media/press-releases/gsk-provides-update-on-herpevac-trial-for-women-evaluating -simplirix-herpes-simplex-vaccine/, accessed October 17, 2019.

38. *US v. Heraeus Electro-Nite Co., LLC*, US District Court for the District of Columbia, Complaint, January 2, 2014.

39. *US v. Heraeus Electro-Nite Co., LLC*, US District Court for the District of Columbia, Final Judgment, April 7, 2014.

40. US Federal Trade Commission, *In the Matter of Nielsen Holdings N.V. and Arbitron Inc.*, File No. 131 0058, Analysis of Agreement Containing Consent Order to Aid Public Comment, September 20, 2013.

41. Ibid.

42. Ibid.

43. See Nielsen, "Audience," available at http://www.nielsen.com/us/en/solutions /measurement/audience.html, accessed October 2, 2019, and Comscore, "Understand and evaluate audiences and advertising everywhere," available at https://www .comscore.com/Products/Audience-Analytics, accessed October 2, 2019.

44. See, e.g., Deveau and Porter (2018).

45. See, e.g., Maddaus (2017). (A Comscore spokesperson said, "We can confirm that there is a dispute regarding interpretation of the 2013 [sic] FTC consent decree, which resulted from Nielsen's acquisition of Arbitron, that provides Comscore with access to certain Nielsen data.")

46. US Federal Trade Commission, *In the Matter of Novartis, AG and GlaxoSmithKline plc*, Docket No. C-4510, Complaint, February 20, 2015.

47. US Federal Trade Commission, *In the Matter of Novartis, AG and GlaxoSmithKline plc*, Analysis of Agreement Containing Consent Orders to Aid Public Comment, File No. 141–0141, February 23, 2015.

48. US Securities and Exchange Commission, Array BioPharma, Inc., Form 10-K, for the fiscal year ending June 30, 2017; and Array BioPharma, "Our pipeline," available at https://www.arraybiopharma.com/product-pipeline, accessed July 23, 2018.

49. See Delrahim (2017).

50. The six cases in Chien (2003) are *In re Roche Holding Ltd.*, 113 F.T.C. 1086 (1990); *In re Institut Merieux S.A.*, 113 F.T.C. 742 (1990); *In re Baxter Int'l Inc.*, 123 F.T.C. 904 (1997); *In re Dow Chem. Co.*, 118 F.T.C. 730 (1994); *In re Ciba-Geigy Ltd.*, 123 F.T.C. 842 (1997); and *In re Eli Lilly & Co.*, 95 F.T.C. 538 (1980). Chien concluded that compulsory licensing obligations caused only Institut Merieux S.A. to reduce its R&D efforts and attributed the reduction to uncertainty caused by a delay in finding a divestiture recipient.

51. Specifically, this refers to patents on the use of herpes simplex virus-thymidine kinase ("HSV-tk") vectors.

52. US Federal Trade Commission, *In the Matter of Ciba-Geigy, Chiron, Sandoz and Novartis*, Docket No. C-3725, Analysis of Proposed Consent Order to Aid Public Comment, December 17, 1996.

53. See Fisher (1994).

54. Sanofi-Aventis Annual Report on Form 20-F, 2009, available at https://www.sanofi .com/en/investors/reports-and-publications/financial-and-csr-reports.

55. Charles Bankhead, "Gene therapy flunks limb ischemia test," *Medpage Today*, June 2, 2011, available at https://www.medpagetoday.com/cardiology/peripheralarterydisease/26814.

56. FDA news release, "FDA approval brings first gene therapy to the United States," August 30, 2017, available at https://www.fda.gov/news-events/press-announcements/fda-approval-brings-first-gene-therapy-united-states.

57. The data reported in table 7.2 is from US Patent and Trademark Office (USPTO) patent searches with "gene therapy" in the patent description/specification. Other search terms produce similar results. For example, USPTO searches with the classification code A61K48/00 ("Medicinal preparations containing genetic material, gene therapy") capture a smaller number of patents, but they have a similar pattern.

58. Serono's annual report for 2003 describes plans to initiate a multicenter, multinational phase III program for Onercept, a recombinant, soluble type I TNF receptor. Serono terminated a phase III study for Onercept in June 2005 after concluding that the risk-benefit ratio was not sufficiently favorable to justify continued development (see https://clinicaltrials.gov/ct2/show/NCT00090129, accessed July 22, 2018). Merck acquired Serono in 2006. Both Serono and Merck describe research efforts for aurorakinase inhibitors, which are related to TNF inhibitors.

59. US Federal Trade Commission, *In the Matter of Flow International Corp.*, File No. 081–0079, Complaint, August 15, 2008. The complaint does not directly allege harm to innovation, although the effect of patent litigation on innovation is an issue in this industry.

60. US Federal Trade Commission, Analysis of the Agreement Containing Consent Order to Aid Public Comment, in the Matter of Flow International Corp., File No. 081–0079.

61. "Waterjet cutting machine buyers' guide," The Fabricator, available at https://www.thefabricator.com/guide/waterjet-cutting-machine, accessed July 23, 2018. The comparison to January 2014 is based on this website listed in archive.org for January 30, 2014, accessed July 29, 2018.

62. *US v. 3D Systems Corporation and DTM Corporation*, Complaint, US District Court for the District of Columbia, June 6, 2001.

63. Hundreds of patents cover RP-related technologies. However, 3D and DTM have major patent portfolios in some specializations. For example, a USPTO search of patents with "stereolithographic printing" in their descriptions returned a total of 533 issued patents, of which 75 were invented by or assigned to 3D Systems.

64. *US v. 3D Systems Corporation and DTM Corporation*, Competitive Impact Statement, US District Court for the District of Columbia, September 4, 2001. The final order also required the merged company to grant software licenses and provide a list of customers (*US v. 3D Systems Corporation and DTM Corporation*, Final Judgment, US District Court for the District of Columbia, April 17, 2002).

65. 3D Systems Corporation, Form 10-K, for the fiscal year ended December 31, 2002, at 19.

66. See, e.g., "Rapid prototyping (RP)—Rapid tooling (RT)—SONY," *Engineers Handbook*, available at http://www.engineershandbook.com/RapidPrototyping/sony.htm, accessed July 10, 2018.

67. A USPTO patent search with the terms "rapid prototyping printing" or "stereolithography" returned 956 successful patent applications in the five-year period from January

1, 1998, to December 31, 2002, of which 89 were filed by 3D Systems. A similar search for the five-year period from January 1, 2003, to December 31, 2007 returned a total of 1,152 successful applications, of which 55 were filed by 3D Systems.

68. 3D Systems Corporation, Annual Reports for 2004 and 2001.

69. "3D printer manufacturers: Who's in the lead?," Fabbaloo, April 16, 2018, available at http://www.fabbaloo.com/blog/2018/4/16/3d-printer-manufacturers-whos-in-the-lead.

70. This section follows the excellent review in Petit (2018).

71. European Commission, *Pasteur Mérieux-Merck*, IV/34.776, October 6, 1994.

72. European Commission, *Glaxo-Wellcome*, IV/M.555, February 28, 1995.

73. US Federal Trade Commission Press Release, "Glaxo plc," June 20, 1995.

74. See CenterWatch, "Zomig (zolmitriptan)," available at https://www.centerwatch .com/drug-information/fda-approved-drugs/drug/347/zomig-zolmitriptan, accessed May 13, 2019.

75. European Commission, *Glaxo Wellcome/SmithKline Beecham*, Case No. COMP/M.1486, August 5, 2000, at ¶ 202.

76. Ibid., at ¶ 214.

77. European Commission, *Astra Zeneca/Novartis*, Case No. COMP/M.1806, July 26, 2000, and US Federal Trade Commission, *In the Matter of Novartis and AstraZeneca*, Docket No. C-3979, Complaint, November 1, 2000.

78. European Commission, *Bayer/Aventis Crop Science*, Case No. COMP/M.2547, April 17, 2002, and US Federal Trade Commission, *In the Matter of Bayer AG and Aventis S.A.*, Docket No. C-4049, Analysis of Proposed Consent Order to Aid Public Comment, May 30, 2002.

79. Both agencies allowed the merger of Seagate and Samsung without conditions and blocked the proposed merger of Western Digital and Viviti Technologies (formerly known as Hitachi Global Storage Technologies). See Statement of the Federal Trade Commission Concerning Western Digital Corporation/Viviti Technologies Ltd. and Seagate Technology LLC/Hard Disk Drive Assets of Samsung Electronics Co. Ltd. and European Commission, *Western Digital Irland/Viviti Technologies*, Case No. COMP/M.6203, November 23, 2011.

80. European Commission, *Medtronic/Covidien*, Case No. COMP/M.7326, November 28, 2014, and US Federal Trade Commission, *In the Matter of Medtronic, Inc. and Covidien plc*, Docket No. C-4503, Decision and Order, January 21, 2015.

81. European Commission Press Release, "Mergers: Commission approves acquisition of Hospira by Pfizer, subject to conditions," August 4, 2015.

82. US Federal Trade Commission, *In the Matter of Pfizer, Inc. and Hospira, Inc.*, Docket No. C-4537, Complaint, August 21, 2015.

83. A biosimilar is a close substitute for its reference biologic, although it is not a generic equivalent.

84. European Commission, *Pfizer/Hospira*, Case No. COMP/M.7559, April 8, 2015, at ¶ 57.

85. Pfizer registered Ixifi for sale in Japan. See Pfizer Form 10-K for the fiscal year ending December 31, 2018.

86. European Commission, *General Electric/Alstom*, Case No. COMP/M.7278, September 8, 2015.

87. The author was a consultant to Dow and DuPont in this transaction.

88. *US et al. v. Dow Chemical Corporation and E.I. DuPont de Nemours and Company*, Competitive Impact Statement, June 15, 2017.

89. European Commission, *Dow/DuPont*, Case No. COMP/M.7932, March 27, 2017.

90. European Commission Press Release, "Mergers: Commission clears merger between Dow and DuPont, subject to conditions." March 27, 2017.

91. European Commission, *Dow/DuPont* (2017), at § 8.6.

92. Ibid., table 67. The HHI measures the sum of the squares of the shares. An HHI of 3,300 corresponds to three firms with equal patent shares.

93. See, e.g., European Commission, *Dow/DuPont* (2017), at ¶ 2016.

94. See, e.g., Delrahim (2004). ("There are important policy reasons to cause us to be cautious when considering a compulsory licensing remedy. The most important of these is the concern that an improperly designed compulsory license can stifle innovation.") However, Delrahim also recognizes that compulsory licensing can be a useful alternative to divestiture to resolve antitrust concerns in some cases.

## Chapter 8

1. The author was the Deputy Assistant Attorney General for Economics at the Antitrust Division of the Department of Justice when the Division initiated its investigation of Microsoft and later consulted with the Division on matters related to its Microsoft complaint.

2. Compare, e.g., Salop and Romaine (1999) with Cass and Hylton (1999).

3. *US v. Microsoft*, US District Court for the District of Columbia, Civil Action No. 98–1232 (TPJ), Complaint (May 18, 1998), at ¶ 6. "IBM-compatible" refers to PCs that are compatible with the IBM basic input/output system (BIOS). The complaint uses the term "Intel-compatible personal computer" because IBM-compatible PCs employed the Intel x86 microprocessor architecture. That term became a misnomer because other microprocessors such as Advanced Micro Devices (AMD) used the Intel architecture, and Intel subsequently supplied microprocessors for Apple and Linux operating systems.

4. Gilbert (1999).

5. IBM and Microsoft initially cooperated to develop OS/2 as a successor to the MS-DOS operating system. The first version of OS/2 launched in 1987 without a graphical user interface (GUI) and met with little success. A second version with a GUI failed because it offered poor support for MS-DOS applications and relatively few device drivers to support hardware other than IBM's. See Evans, Nichols, and Reddy (2002).

6. OS/2 Warp was not without its disadvantages, including larger memory requirements than Windows and IBM marketing missteps. See Swedin (2009). The fact that most Intel-compatible PCs shipped in the 1990s with Windows preinstalled was another factor that worked against the success of OS/2. See Evans et al. (2002).

7. "Memo from Bill Gates to executive staff and direct reports," The Internet Tidal Wave, May 26, 1995, available at https://www.justice.gov/sites/default/files/atr/legacy/2006/03/03/20.pdf.

8. The term "Java" refers to a set of technologies that include (1) a programming language; (2) a set of programs written in that language, called the Java class libraries, which expose APIs; (3) a compiler, which translates code written by a developer into "bytecode" (low-level code); and (4) a Java virtual machine (JVM), which translates bytecode into instructions to the operating system. Programs calling upon the Java APIs will run on any machine with a "Java runtime environment"; that is, Java class libraries and a JVM. (See *US et al. v. Microsoft*, US District Court for the District of Columbia, Civil Action No. 98–1232 [TPJ], Findings of Fact [November 5, 1999], at ¶ 73.) Mentions of Java in this chapter do not refer to JavaScript, which is used for interactive web pages.

9. "Memo from Bill Gates to executive staff and direct reports," The Internet Tidal Wave, May 26, 1995, available at https://www.justice.gov/sites/default/files/atr/legacy/2006/03/03/20.pdf. A client is a piece of hardware or software, such as a web browser, that interacts with a network server.

10. Cusumano and Yoffie (1998) at 40.

11. Katz and Shapiro (1985).

12. U.S v. Microsoft, Complaint at ¶ 14, 70–73.

13. Ibid. at ¶ 16.

14. A second browser war broke out in the early 2000s, after Netscape spun off Navigator into the open-source Mozilla project, which developed the Firefox product, and Google followed with its Chrome browser. See, e.g., Steve Lovelace, "The second browser war," July 31, 2015, available at http://steve-lovelace.com/the-second-browser-war/.

15. *US v. Microsoft*, Complaint at ¶ 64.

16. European Commission Press Release, "Commission examines the impact of Windows 2000 on competition," February 10, 2000. ("The allegations which we have now decided to examine more closely centre on Microsoft leveraging its dominance from one market (PC operating systems) onto other markets, whereas in the US the main thrust of the proceedings seems to revolve around Microsoft protecting its dominance on the market for PC operating systems.") Compared to client PCs, servers are more powerful machines that can support multiple users and have the capacity to store large amounts of data and files.

17. Rubinfeld (2008) provides a useful summary of the allegations in *US v. Microsoft* and Microsoft's responses. Gavil and First (2014) provide a detailed description of the US antitrust case and several other related cases brought against Microsoft by US and European antitrust enforcers.

18. *US v. Microsoft*, Findings of Fact at ¶ 94–114 and ¶ 345–356.

19. Ibid., Section VI; and *US et al. v. Microsoft*, US District Court for the District of Columbia, Civil Action No. 98–1232 (TPJ), Conclusions of Law (April 3, 2000), at 42–43.

20. *US v. Microsoft*, Complaint at ¶ 37.

21. *US v. Microsoft*, Conclusions of Law, 87 F. Supp. 2d 30, 38.

22. Ibid. at 39.

23. *US v. Microsoft*, Findings of Fact at ¶ 77.

24. The district court found that Microsoft's contracts with OEMs, ISVs, and ICPs did not satisfy the threshold requirements for unlawful exclusive dealing, in violation of Section 1 of the Sherman Act.

25. *US v. Microsoft*, Conclusions of Law, 87 F. Supp. 2d 30, 43.

26. *US v. Microsoft*, Final Judgment (November 12, 2002).

27. Courts have traditionally held tied sales to be per se unlawful when (1) two separate products or services are involved, (2) the sale or agreement to sell one is conditioned on the purchase of the other, (3) the seller has sufficient economic power in the market for the tying product to enable it to restrain trade in the market for the tied product, and (4) a not-insubstantial amount of interstate commerce in the tied product is affected. (See US Department of Justice and Federal Trade Commission, 2007, 105–106.)

28. *US v. Microsoft*, Court of Appeals, 253 F.3d 34, 85 (June 28, 2001).

29. Ibid. at 53.

30. Ibid. at 91.

31. One plaintiff state held out for additional sanctions, which the court rejected in June 2004. *Massachusetts v. Microsoft Corp.*, 373 F. 3d 1199 (D.C. Cir., 2004).

32. See Hesse (2009) for a detailed discussion of the settlement terms.

33. Commission Decision relating to a proceeding pursuant to Article 82 of the EC Treaty and Article 54 of the EEA Agreement against Microsoft Corporation (Case COMP/ C-3/37.792—*Microsoft*), May 24, 2004. Affirmed by Judgment of the Court of First Instance, Case T-201/04, September 17, 2007.

34. Ayres and Nalebuff (2005) applaud the remedy, notwithstanding the few sales of Windows with WMP, because it addressed the alleged unlawful tie and did not impose costs on consumers.

35. Commission Decision relating to a proceeding pursuant to Article 82 of the EC Treaty (Case COMP/C-3/37.792—*Microsoft*), May 24, 2004 at ¶ 784.

36. See Kühn and Van Reenen (2009) for a useful discussion of the interoperability issues in the EC investigation of Microsoft.

37. Commission Decision relating to a proceeding pursuant to Article 82 of the EC Treaty (Case COMP/C-3/37.792—*Microsoft*), March 24, 2004 at ¶ 999.

38. Ibid. at ¶ 1007.

39. European Commission, "Antitrust: Commission initiates formal investigations against Microsoft in two cases of suspected abuse of dominant market position," Memo/08/19, January 14, 2008.

40. European Commission, "Antitrust: Commission market tests Microsoft's proposal to ensure consumer choice of web browsers; welcomes further improvements in field of interoperability," Memo/09/439, October 7, 2009.

41. Commission Decision, Case COMP/C-3/39.530—*Microsoft (tying)*, December 16, 2009.

42. Ibid.

43. Wikipedia, "Usage share of web browsers (citing several historical data sources)," available at https://en.wikipedia.org/wiki/Usage_share_of_web_browsers, accessed April 16, 2019.

44. 540 US 398, 406 (2004).

45. Reynolds and Best (2012) review EC jurisprudence on refusals to deal by dominant firms. Chapter 9 addresses allegations of anticompetitive product design in the context of the Google Search case, where the US and European antitrust enforcers reached different conclusions.

46. Commission Decision relating to a proceeding pursuant to Article 82 of the EC Treaty and Article 54 of the EEA Agreement against Microsoft Corporation (Case COMP/C--3/37.792—*Microsoft*), May 24, 2004.

47. Council Directive on the legal protection of computer programs (91/250/EEC), May 14, 1991.

48. European Commission Decision, *Sea Containers Sealink/Stena*, OJ 1994 L15/8 (December 21, 1993), quoted in Vickers (2010).

49. Commission decision of December 21, 1988, *Magill TV Guide/ITP, BBC and RTE* ([1989] OJ L78/43) and Case C-418/01 *IMS Health GmbH & Co. OHG v. NDC Health GmbH &Co KG*, judgment of April 29, 2004.

50. *US v. Microsoft*, Court of Appeals, 253 F.3d 34, 11–12.

51. Ibid. at 57–59, 64–66.

52. On remand, the court would have allowed the plaintiffs to argue that the anticompetitive effects in the browser market from the Windows 98 override of a consumer's choice of default web browser outweighed Microsoft's claimed efficiency benefits for the operating system, but the Settlement preempted that investigation.

53. Segal and Whinston (2007) explore these trade-offs with a model of sequential innovation in which antitrust law determines the extent to which incumbents can profit by excluding entrants. See also Baker (2007, 2016).

54. Hylton and Lin (2013) argue that innovations create consumer benefits and should not be deterred by antitrust enforcement. However, they do not consider how monopoly conduct can harm rival innovation or slow the progress of future innovation.

55. Commission Decision relating to a proceeding pursuant to Article 82 of the EC Treaty and Article 54 of the EEA Agreement against Microsoft Corporation (Case COMP/C--3/37.792—*Microsoft*), May 24, 2004.

56. European Commission, "Antitrust: Commission imposes € 899 million penalty on Microsoft for non-compliance with March 2004 decision," IP/08/318, February 27, 2008.

57. *New York ex. rel. v. Microsoft*, California Group's Report on Remedial Effectiveness, Civil Action 98-1233 (CKK) (D.D.C. September 11, 2007).

58. *US v. Microsoft*, Joint Status Report on Microsoft's Compliance with the Final Judgments, (US D.D.C) Civil Action No. 98–1232 (CKK), April 27, 2011.

59. The Chrome, Firefox, and Safari browsers do not offer a Java plugin applet for their current versions, although the browsers can access websites that support Java applications.

See https://java.com/en/download/help/enable_browser.xml, accessed October 16, 2019.

60. Google used Java to code the major server-side functions of Google Docs. See Jonathan Strickland, "How Google Docs works," available at https://computer.howstuffworks.com/internet/basics/google-docs5.htm, accessed April 16, 2019.

61. See, e.g., Carlton and Waldman (2002) and Nalebuff (2004).

62. George Stigler was the first to articulate the benefits from mixed bundling (Stigler, 1963), followed by Adams and Yellen (1976). Janet Yellen went on to chair the Board of Governors of the US Federal Reserve from 2014–2018.

63. See, e.g., Evans and Salinger (2005).

64. An additional argument is that a commitment to offer products only as a bundle can make a firm a more formidable competitor, which can make new entry that requires an irreversible investment more difficult. See Whinston (1990, 2001). However, this argument does not have much relevance for the tying allegation in the Microsoft case because Netscape was an established competitor.

65. See, e.g., *US et al. v. Microsoft*, US District Court for the District of Columbia, Civil Action No. 98-1232 (TPJ), Declaration of Carl Shapiro (April 28, 2000) and Declaration of Paul M. Romer (April 27, 2000).

66. Ordover, Saloner, and Salop (1990) and Allain, Chambolle, and Rey (2011, 2016) describe the incentives of an integrated supplier to disadvantage rivals. Carlton and Waldman (2002) and Choi (2004) show that tying (which is closely related to integration) can reduce rival innovation incentives.

67. See, e.g., Bresnahan (2001), Ayres and Nalebuff (2005), First and Gavil (2006), Hovenkamp (2008a), and Shapiro (2009). Herbert Hovenkamp wrote that "the Microsoft case may prove to be one of the great debacles in the history of public antitrust enforcement, snatching defeat from the jaws of victory." Hovenkamp (2008a) at 298.

68. The innovation incentive is not as general as the Cournot complements effect for prices because consumers who purchase operating systems and applications from an integrated supplier could have a low value for improvements at the lower prices that prevail under integrated supply. See the appendix to Farrell and Katz (2000) and the discussion in chapter 2.

69. Shelanski and Sidak (2001) at 30. ("One can easily imagine, to take only one example, that the meaning of 'middleware' would be thoroughly litigated by interested parties, just as the meaning of 'information services' was thoroughly litigated under the Modification of Final Judgment [the AT&T consent decree].")

70. See Farrell and Katz (2000).

71. See Williamson (1979, 1983).

72. According to Statcounter, more than two-thirds of desktop internet use was with Chrome in 2019, compared to less than 15 percent with Microsoft's Internet Explorer and its replacement, Microsoft Edge. See http://gs.statcounter.com/browser-market-share/desktop/worldwide, accessed October 15, 2019.

73. Heiner (2012) at 340.

74. *US v. Microsoft*, Findings of Fact at ¶ 407.

75. *US v. Microsoft*, Court of Appeals, 253 F.3d 34, 61 (emphasis added).

76. See, e.g., Shapiro (1999).

77. Hovenkamp (2008b) observes that a corollary of the premise that innovation contributes more to economic growth than price competition is that a restraint on innovation can do much more harm.

78. See Farrell and Katz (2000) and Elhauge (2009).

79. Hesse (2009) at 865–868.

## Chapter 9

1. The author consulted with the FTC in its Google Search investigation. This chapter, which is based in part on Gilbert (2018b), does not address other search-related allegations against Google, including the appropriation of information from websites and restrictions on the use of programs to manage advertising. In a separate case, the EC fined Google €1.49 billion for abusing its market dominance by imposing clauses in contracts with third-party websites which prevented rivals from placing their search ads on these websites. (European Commission Press Release, "Antitrust: Commission fines Google €1.49 billion for abusive practices in online advertising," Brussels, March, 20, 2019.) In another case, the EC fined Google €4.34 billion and ordered the company to end restrictions related to the licensing of its Android mobile operating system. (European Commission Press Release, "Antitrust: Commission fines Google €4.34 billion for illegal practices regarding Android mobile devices to strengthen dominance of Google's search engine," Brussels, July 18, 2018.) I briefly discuss the Google Android case in chapter 10. I do not include these cases in part for space reasons, and in part because they raise issues related to tying and exclusive dealing covered in the discussion of the Microsoft antitrust litigation in chapter 8.

2. See Google Ads Help, "Ad Rank," available at https://support.google.com/google -ads/answer/1752122, accessed February 14, 2019.

3. See, e.g., "From ten blue links to integrated information platform," in Crane (2012).

4. Baye, de los Santos, and Wildenbeest (2016).

5. Google Algorithm Update History, available at https://moz.com/google-algorithm -change, accessed October 23, 2019.

6. The update is named for the Google engineer Navneet Panda, who developed the technology embodied in the algorithm. See Brafton, "Google Panda," available at https://www.brafton.com/glossary/google-panda/, accessed January 25, 2019.

7. A Google blogpost stated: "This [Panda] update is designed to reduce rankings for low-quality sites—sites which are low-value add for users, copy content from other websites or sites that are just not very useful. At the same time, it will provide better rankings for high-quality sites—sites with original content and information such as research, in-depth reports, thoughtful analysis and so on" (https://googleblog.blogspot .com/2011/02/finding-more-high-quality-sites-in.html, accessed January 25, 2019).

8. E-mail from Google employee, quoted in European Commission, *Google Search (Shopping)*, AT. 39740, Decision, June 27, 2017, at ¶ 382 ("EC Google Shopping Decision").

9. These statistics come from Statcounter (http://gs.statcounter.com/search-engine -market-share/, accessed January 25, 2019). Query shares vary according to device,

location, and how they are measured. Comscore reported that Google accounted for 64 percent of direct search queries by US users in February 2016 ("Comscore releases February 2016 US desktop search engine rankings," March 16, 2016, available at https://www .comscore.com/Insights/Rankings/comScore-Releases-February-2016-US-Desktop -Search-Engine-Rankings).

10. "If you do not like the answer that Google search provides you can switch to another engine with literally one click...," Testimony of Eric Schmidt, Executive Chairman, Google Inc., Before the Senate Committee on the Judiciary Subcommittee on Antitrust, Competition Policy, and Consumer Rights, September 21, 2011, at 6.

11. See, e.g., Manne and Wright (2011) and Ratliff and Rubinfeld (2014). More generally, the ability to raise prices or exclude competition on one side of a two-sided market depends on reactions by firms or consumers on the other side. See, e.g., Evans and Noel (2005) and Ratliff and Rubinfeld (2010).

12. See, e.g., Patterson (2013). See Darby and Karni (1973) for a discussion of credence goods.

13. Google processes about 3.5 billion search queries every day, which adds up to more than a trillion per year; see Google Search Statistics, available at http://www.internetlivestats .com/google-search-statistics/, accessed January 25, 2019. As of the end of 2018, the total number of searches on DuckDuckGo was 26 billion; see https://duckduckgo.com/about, accessed January 25, 2019.

14. See also Luca et al. (2015).

15. See, e.g., Allain, Chambolle, and Rey (2011, 2016).

16. Langford (2013) and Stucke and Ezrachi (2016).

17. *US v. Google and ITA Software*, Complaint, US District Court for the District of Columbia, Case: 1:11-cv-00688 (April 8, 2011).

18. The FTC addressed other Google search-related conduct, but it accepted Google's voluntary agreement to change the challenged practices and did not require a formal consent decree (US Federal Trade Commission, 2013a).

19. Ibid.

20. European Commission Press Release, "Antitrust: Commission fines Google €2.42 billion for abusing dominance as search engine by giving illegal advantage to own comparison shopping service—Factsheet," Brussels, June 27, 2017, available at http://europa .eu/rapid/press-release_MEMO-17-1785_en.htm.

21. EC Google Shopping Decision at ¶ 154.

22. Ibid. at ¶ 271.

23. Ibid. at ¶ 341.

24. Ibid. See also European Commission, Summary of Commission Decision of 27 June 2017 Relating to a Proceeding under Article 102 of the Treaty on the Functioning of the European Union and Article 54 of the EEA Agreement, Case AT.39740—Google Search (Shopping).

25. EC Google Shopping Decision at § 7.5.

26. Ibid. at ¶ 333.

27. Consolidated version of the Treaty on the Functioning of the European Union—Rules Applying to Undertakings—Article 102 (ex Article 82 TEC). See also Vickers (2005).

28. EC Google Shopping Decision at ¶ 593–596.

29. Ibid. at ¶ 538.

30. Ibid. at ¶ 144.

31. European Commission Press Release, "Antitrust: Commission fines Google €1.49 billion for abusive practices in online advertising," Brussels, March 20, 2019.

32. Ibid.

33. There are exceptions. The US Court of Appeals for the Federal Circuit held that a design change whose intent was to exclude rivals was anticompetitive notwithstanding some evidence that the change was an improvement. See *C.R. Bard, Inc. v. M3 Systems, Inc.*, US Court of Appeals for the Federal Cir. (September 30, 1998). The precedential significance of this case is limited because the anticompetitive allegations were not thoroughly briefed on appeal. The case diverges from traditional US antitrust jurisprudence, which focuses on competitive effects, not intent.

34. *US v. Microsoft Corp.*, 253 F.3d 34, 64 (June 28, 2001).

35. *Berkey Photo v. Eastman Kodak Co.*, US Court of Appeals for the 2nd Circuit, 603 F.2d 263, 286 (June 25, 1979) (emphasis added).

36. In 2010, pharmaceutical companies spent $27.7 billion on promotion and direct-to-consumer advertising, an amount equal to about 9 percent of their sales (Kornfield, Donohoe, Berndt, and Alexander, 2013).

37. Carrier and Shadowen (2016, 2018) make this argument.

38. Jones and Williams (1998).

39. See, e.g., Newman (2012).

40. See, e.g., *Cal. Computer Prods. v. IBM Corp.*, 613 F.2d 727 (1979) and *Memorex Corp. v. IBM Corp.*, 636 F.2d 1188 (1980) (design changes lowered costs and increased performance). In *Transamerica Computer Co. v. IBM Corp.*, 698 F.2d 1377 (9th Cir. 1983), a district court held that a design change was anticompetitive but the plaintiff did not suffer antitrust injury.

41. *Verizon Communications, Inc. v. Law Offices of Curtis V. Trinko*, US Supreme Court (January 13, 2004). The court did not mention the qualification that the challenged conduct in Trinko occurred in a regulated industry.

42. *In re Apple iPod iTunes Antitrust Litigation*, US District Court for the Northern District of California (May 19, 2011).

43. *In re Apple iPod iTunes Antitrust Litigation*, Verdict Form Re Genuine Product Improvement (N.D. Cal. 2014).

44. See Werden (2006) at 419, which cites Areeda and Hovenkamp (2002).

45. *Allied Orthopedic Appliances, Inc. v. Tyco Health Care Group LP*, 592 F.3d 991, 999 (January 6, 2010) (citations omitted).

46. Ordover and Willig (1981).

47. See Werden (2006) and Melamed (2006).

48. See, e.g., Elhauge (2003).

49. Steve Jobs, *BusinessWeek*, October 12, 2004.

50. US Federal Trade Commission, *In the Matter of Intel Corporation*, Docket No. 9341, Complaint (December 16, 2009).

51. US Federal Trade Commission, *In the Matter of Intel Corporation*, Docket No. 9341, Decision and Order (October 29, 2010). The decree requires only a showing of actual benefits to Intel from a design change; it does not require a balancing of costs and benefits, although the FTC disclosed that a balancing would be appropriate if the design change were challenged under the antitrust laws. See *In the Matter of Intel Corporation*, Docket No. 9341, Analysis of Proposed Consent Order to Aid Public Comment.

52. See, e.g., Werden (2006) at 416.

53. See, e.g., Salop (2006).

54. Vickers (2005).

55. Hovenkamp (2013).

56. Gilbert (2007).

57. European Commission—Fact Sheet, "Antitrust: Commission fines Google €2.42 billion for abusing dominance as search engine by giving illegal advantage to own comparison shopping service," Brussels, June 27, 2017.

58. See, e.g., Creighton and Jacobson (2012). These authors argue that the Supreme Court opinion in *Verizon v. Trinko* is a departure from prior court decisions that imposed duties on a dominant firm to assist its rivals.

59. *Lorain Journal Co. v. United States*, 342 U.S. 143 (1951).

60. *Otter Tail Power Co. v. United States*, 410 U.S. 366 (1973). A subsequent case that is often cited in the context of allegations of refusals to deal is *Aspen Skiing Co. v. Aspen Highlands Skiing Corp.*, 472 U.S. 585 (1985). That case premised antitrust liability for a refusal to deal on the termination of a profitable prior course of dealing, which is irrelevant for some types of conduct that have anticompetitive consequences. See, e.g. Creighton and Jacobson (2012).

## Chapter 10

1. One estimate is that a modern laptop computer embodies more than 250 technologies defined by standards. See Biddle (2018).

2. This chapter is based in part on Gilbert (2014).

3. See, e.g., Shapiro and Varian (1999a, 1999b).

4. See, e.g., Farrell and Saloner (1985, 1986).

5. Simcoe (2012) shows that commercialization of the internet caused an increase in strategic maneuvering within the IETF and a slowdown in committee decision-making regarding internet standards.

6. See, e.g., Anton and Yao (1995).

7. In 1997, Sun Microsystems applied for and won approval to submit Java to the International Committee for Information Technology Standardization. Sun ultimately abandoned its efforts to certify Java as a public standard over disputes about intellectual property protection. See, e.g., Garud et al. (2002).

8. European Commission, *Google Android,* Case AT.40099, Commission Decision, 18 July 2018 at Section 12.6.1.

9. Ibid.

10. European Commission Press Release, "Antitrust: Commission fines Google €4.34 billion for illegal practices regarding Android mobile devices to strengthen dominance of Google's search engine," July 18, 2018.

11. ANSI Patent Policy, revised 2016, available at https://share.ansi.org/Shared%20 Documents/Standards%20Activities/American%20National%20Standards /Procedures,%20Guides,%20and%20Forms/ANSI%20Patent%20Policy%202016.pdf.

12. Goodman and Myers (2005). WCDMA and CDMA2000 are third-generation cellular communications technologies. The initials WCDMA stand for "Wideband Code Division Multiple Access." A patent family includes all the patents registered by one or more common inventors in different countries to protect an invention.

13. See Williamson (1979) and Farrell, Hayes, Shapiro, and Sullivan (2007) for a discussion of the economics of holdup in the context of standard-setting.

14. See Shapiro (2001).

15. See Lemley and Shapiro (2007). Royalty-stacking is an example of the Cournot complements effect discussed in chapter 2.

16. ANSI Patent Policy.

17. See, e.g., Goodman and Myers (2005) and Maskus and Merrill (2013).

18. US Federal Trade Commission, *In the Matter of Dell Computer Corporation,* Complaint, May 20, 1996.

19. US Federal Trade Commission, *In the Matter of Dell Computer Corporation,* Statement of the Commission, June 17, 1996.

20. The author testified on behalf of private plaintiffs adverse to Rambus in a related case.

21. US Federal Trade Commission, *In the Matter of Rambus Inc.,* Docket No. 9302, Opinion of the Commission (August 2, 2006).

22. *Rambus Inc. v. FTC,* US Court of Appeals for the District of Columbia, (April 22, 2008), 522 F.3d 456.

23. 522 F.3d 456, 456 (emphasis in original). "SSO" stands for "standards setting organization," which is another term for "standards development organization."

24. The court of appeals in *Rambus* did not directly address whether Rambus might have violated Section 5 of the Federal Trade Commission Act, although it noted that JEDEC's vague disclosure requirements would cast doubt on such a finding.

25. 522 F.3d 456, 465. Italics in original.

26. See, e.g., Galetovic, Haber, and Levine (2015).

27. Galetovic, Haber, and Levine (2015) also claim that holdup is not an economic concern because court decisions that reduce the excessive power of SEP holders have not accelerated innovation in SEP-reliant industries. However, there is not sufficient data to estimate such innovation effects accurately.

28. See, e.g., Carey and Culley (2018).

29. *Allied Tube & Conduit Corp. v. Indian Head*, US Supreme Court (June 13, 1988).

30. Simcoe (2012).

31. Arrow (1950).

32. Many such allegations survive a motion to dismiss or summary judgment, but then are dismissed on the merits after a finding that standards were developed in a process that followed these principles. Examples include *Addamax v. Open Software Foundation*, 888 F. Supp. 274 (D. Mass. 1995), aff'd, 152 F.3d 48 (1st Cir. 1998) (not unlawful to exclude technology from Unix OSF/1); and *Golden Bridge Tech., Inc. v. Nokia, Inc.* (E.D. Tex., Sept. 10, 2007), aff'd *Golden Bridge Tech., Inc. v. Motorola, Inc.*, 547 F.3d 266 (5$^{th}$ Cir. 2008) (not unlawful to exclude technology from 3G cellular standards).

33. *GSI Tech., Inc. v. Cypress Semiconductor Corp.*, US Dist. LEXIS 9378 (January 27, 2015). The case ended after the parties reached an undisclosed settlement. Cypress Semiconductor, 2015 Annual Report at 104.

34. See, e.g, *Microsoft Corporation v. Motorola, Inc., et al.*, 2013 US Dist. LEXIS 60233 (W.D. Wash., April 25, 2013); *In re Innovatio IP Ventures, LLC Patent Litig.*, 2013 US Dist. LEXIS 144061 (N.D. Ill., October 3, 2013); and *Commonwealth Sci. & Indus. Research Organisation v. Cisco Sys.* (Fed. Cir., December 3, 2015).

35. VITA Standards Organization, *VSO Policies and Procedures*, September 1, 2015, Revision 2.8.

36. Letter from Thomas O. Barnett, Assistant Attorney General, to Robert A. Skitol, October 30, 2006, available at https://www.justice.gov/atr/response-vmebus-international-trade -association-vitas-request-business-review-letter.

37. Letter from Renata B. Hesse, acting assistant attorney general, to Michael A. Lindsay, February 2, 2015, available at https://www.justice.gov/atr/response-institute-electrical -and-electronics-engineers-incorporated.

38. See, e.g., Lerner and Tirole (2006) and Chiao, Lerner and Tirole (2017).

39. Contreras and Gilbert (2015).

40. Delrahim (2018a, 2018b).

41. See, e.g., *In the Matter of Negotiated Data Solutions LLC*, Complaint (September 23, 2008) and *In the Matter of Robert Bosch Gmbh*, Complaint (November 26, 2012). Issues associated with FRAND licensing agreements are examined at length in Maskus and Merrill (2013).

42. Gilbert (2018a).

43. See Gilbert (2011) and Carlton and Shampine (2013).

44. Contreras and Gilbert (2015) and Lee and Melamed (2016).

45. Ibid.

46. See, e.g, *Ericsson, Inc. v. D-Link Sys.* (Fed. Cir. December 4, 2014), 773 F.3d 1201, 1226, 1231.

47. Ratliff and Rubinfeld (2013) propose a scheme to determine when a patent owner may reasonably request an injunction or exclusion order. Geradin and Rato (2007) argue that a FRAND commitment should not preclude seeking an injunction when good faith negotiations have failed.

48. Merges and Kuhn (2009) at 4.

# References

Adams, William J. and Janet L. Yellen (1976). Commodity bundling and the burden of monopoly. *Quarterly Journal of Economics*, 90(3): 475–498.

Aghion, Philippe, Nick Bloom, Richard Blundell, Rachel Griffith, and Peter Howitt (2005). Competition and innovation: An inverted-U relationship. *Quarterly Journal of Economics*, 120(2): 701–728.

Aghion, Philippe, Richard Blundell, Rachel Griffith, Peter Howitt, and Susanne Prantl (2009). The effects of entry on incumbent innovation and productivity. *Review of Economics and Statistics*, 91(1): 20–32.

Aghion, Philippe, Christopher Harris, Peter Howitt, and John Vickers (2001). Competition, imitation and growth with step-by-step innovation. *Review of Economic Studies*, 68(3): 467–492.

Aghion, Philippe, Christopher Harris, and John Vickers (1997). Competition and growth with step-by-step innovation: An example. *European Economic Review*, 41(3–5): 771–782.

Aghion, Philippe and Jean Tirole (1994). The management of innovation. *Quarterly Journal of Economics*, 109(4): 1185–1209.

Allain, Marie-Laure, Claire Chambolle, and Patrick Rey (2011). Vertical integration, information and foreclosure. Toulouse School of Economics, Working paper, November 25.

Allain, Marie-Laure, Claire Chambolle, and Patrick Rey (2016). Vertical integration as a source of hold-up. *Review of Economic Studies*, 83: 1–25.

Anand, Geeta (2006). *The Cure: How a How a Father Raised $100 Million—and Bucked the Medical Establishment—in a Quest to Save His Children*. HarperCollins.

Antitrust Subcommittee (1959). *Consent Decree Program of the Department of Justice*. Committee on the Judiciary, House of Representatives. Eighty-Sixth Congress, First Session.

Anton, James J. and Dennis A. Yao (1995). Standard-setting consortia, antitrust, and high-technology industries. *Antitrust Law Journal*, 64: 247–265.

Armstrong, Mark (2006). Competition in two-sided markets. *Rand Journal of Economics*, 37(3): 668–691.

Arrow, Kenneth J. (1950). A difficulty in the concept of social welfare. *Journal of Political Economy*, 58: 328–346.

Arrow, Kenneth J. (1962). Economic welfare and the allocation of resources for invention. In Richard R. Nelson (Ed.), *The Rate and Direction of Inventive Activity* (pp. 609–626). Princeton University Press.

Aurelian, Laure (2004). Herpes simplex virus type 2 vaccines: New ground for optimism? *Clinical and Diagnostic Laboratory Immunology*, 11(3): 437–445.

Autor, David, David Dorn, Gordon H. Hanson, Gary Pisano, and Pian Shu (2020). Foreign competition and domestic innovation: Evidence from US patents. *American Economic Review*, forthcoming.

Autor, David, David Dorn, Lawrence F. Katz, Christina Patterson, and John Van Reenen (2017). The fall of the labor share and the rise of superstar firms. Available at https://economics.mit.edu/files/12979.

Ayres, Ian and Barry Nalebuff (2005). Going soft on Microsoft? The EU's antitrust case and remedy. *The Economists' Voice*, 2(2): 1–10.

Baker, Jonathan B. (2007). Beyond Schumpeter vs. Arrow: How antitrust fosters innovation. *Antitrust Law Journal*, 74(3): 575–602.

Baker, Jonathan B. (2016). Evaluating appropriability defenses for the exclusionary conduct of dominant firms in innovative industries. *Antitrust Law Journal*, 80: 431–461.

Baker, Jonathan B. and Carl Shapiro (2008). Reinvigorating horizontal merger enforcement. In Robert Pitofsky (Ed.), *How the Chicago School Overshot the Mark: The Effect of Conservative Economic Analysis on US Antitrust* (pp. 235–238). Oxford University Press.

Baldwin, William L. and John T. Scott (1987). Market structure and technological change. In J. Lesourne and H. Sonnenschein (Eds.), *Fundamentals of Pure and Applied Economics*. Harwood Academic Publishers.

Bankhead, Charles (2011). Gene therapy flunks limb ischemia test. *Medpage Today*, June 2. Available at https://www.medpagetoday.com/cardiology/peripheralarterydisease/26814.

Baten, Joerg, Nicola Bianchid, and Petra Moser (2017). Compulsory licensing and innovation—historical evidence from German patents after WWI. *Journal of Development Economics*, 126: 231–242.

Baxter, William F. (1984–1985). The definition and measurement of market power in industries characterized by rapidly developing and changing technologies. *Antitrust Law Journal*, 53: 717–732.

Baye, Michael R., Barbur de los Santos, and Matthijs R. Wildenbeest (2016). Search engine optimization: What drives organic traffic to retail sites? *Journal of Economics & Management Strategy*, 25(1): 6–31.

Bessen, James and Eric Maskin (2009). Sequential innovation, patents, and imitation. *RAND Journal of Economics*, 40(4): 611–635.

Biddle, C. Bradford (2018). No standard for standards: Understanding the ICT standards-development ecosystem. In Jorge L. Contreras (Ed.), *The Cambridge Handbook of Technical Standardization Law: Competition, Antitrust, and Patents* (pp. 17–28). Cambridge University Press.

Biotech Industry Report (2015). *Beyond Borders: Reaching New Heights.*

Blonigen, Bruce A. and Justin R. Pierce (2016). Evidence of the effects of mergers on market power and efficiency. NBER Working Paper no. 22750, October.

Bloom, Nicholas, Mirko Draca, and John Van Reenen (2016). Trade-induced technical change? The impact of Chinese imports on innovation, IT, and productivity. *Review of Economic Studies*, 83(1): 87–117.

Blundell, Richard, Rachel Griffith, and John Van Reenen (1999). Market share, market value and innovation in a panel of British manufacturing firms. *Review of Economic Studies*, 66(3): 529–554.

Bourreau, Marc and Bruno Jullien (2018). Mergers, investment, and demand expansion. *Economic Letters*, 167(C): 136–141.

Bourreau, Marc, Bruno Jullien, and Yassine Lefouili (2018). Mergers and demand-enhancing innovation. Toulouse School of Economics, Working Paper No. 18–907, March.

Bresnahan, Timothy F. (1985). Post-entry competition in the plain paper copier market. *American Economic Review, Papers and Proceedings*, 75(2): 15–19.

Bresnahan, Timothy F. (2001). A remedy that falls short of restoring competition. *Antitrust*, Fall: 67–71.

Brodley, Joseph F. (1987). The economic goals of antitrust: Efficiency, consumer welfare, and technological progress. *New York University Law Review*, 62: 1020–1053.

Bulow, Jeremy I., John D. Geanakoplos, and Paul D. Klemperer (1985). Multimarket oligopoly: Strategic substitutes and complements. *Journal of Political Economy*, 93(3): 488–511.

Cabral, Luís (1994). Bias in market R&D programs. *International Journal of Industrial Organization*, 12: 533–547.

Cabral, Luís and David Salant (2014). Evolving technologies and standards regulation. *International Journal of Industrial Organization*, 36: 48–56.

Carey, George S. and Daniel P. Culley (2018). Concerted action in standard-setting. In Jorge L. Contreras (Ed.), *The Cambridge Handbook of Technical Standardization Law: Competition, Antitrust, and Patents* (pp. 61–77). Cambridge University Press.

Carlton, Dennis W. and Robert H. Gertner (2003). Intellectual property, antitrust, and strategic behavior. In Adam B. Jaffe, Josh Lerner, and Scott Stern (Eds.), *Innovation Policy and the Economy* (Vol. 3) (pp. 29–59). MIT Press.

Carlton, Dennis W. and Allan L. Shampine (2013). An economic interpretation of FRAND. *Journal of Competition Law and Economics*, 9(3): 531–552.

Carlton, Dennis W. and Michael Waldman (2002). The strategic use of tying to preserve and create market power in evolving industries. *RAND Journal of Economics*, 33(2): 194–220.

Carrier, Michael A. (2008). Two puzzles resolved: of the Schumpeter-Arrow stalemate and pharmaceutical innovation markets. *Iowa Law Review*, 93(2): 393–450.

Carrier, Michael A. and Steve D. Shadowen (2016). Product hopping: A new framework. *Notre Dame Law Review*, 92(1): 167–230.

Carrier, Michael A. and Steve D. Shadowen (2018). A non-coercive economic approach to product hopping. *Antitrust*, 33(1): 102–107.

Cartwright, Heather and Taskin Ahmed (2016). *IMS PharmaDeals: Review of 2016*. QuintilesIMS.

Cass, Ronald A. and Keith N. Hylton (1999). Preserving competition: Economic analysis, legal standards, and Microsoft. *George Mason Law Review*, 8: 1–40.

Cassiman, Bruno, Massimo G. Colombo, Paola Garrone, and Reinhilde Veugelers (2005). The impact of M&A on the R&D process: An empirical analysis of the role of technological- and market-relatedness. *Research Policy*, 34(2): 195–220.

Chang, Howard F. (1995). Patent scope, antitrust policy, and cumulative innovation. *RAND Journal of Economics*, 26(1): 34–57.

Chen, Yongmin and Marius Schwartz (2013). Product innovation incentives: Monopoly vs. competition. *Journal of Economics and Management Strategy*, 22(3): 513–528.

Chiao, Benjamin, Josh Lerner, and Jean Tirole (2017). The rules of standard-setting organizations: An empirical analysis. *RAND Journal of Economics*, 38(4): 905–930.

Chien, Colleen (2003). Cheap drugs at what price to innovation: Does the compulsory licensing of pharmaceuticals hurt innovation? *Berkeley Technology Law Journal*, 18: 853–907.

Choi, Jay Pil (2004). Tying and innovation: A dynamic analysis of tying arrangements. *Economic Journal*, 114: 83–101.

Christensen, Clayton (1997). *The Innovator's Dilemma*. Harvard Business School Press.

Church, Jeffrey and Neil Gandal (1993). Complementary network externalities and technological adoption. *International Journal of Industrial Organization*, 11: 239–260.

Clark, Don and Ezequiel Minaya (2016). Lam Research, KLA-Tencor call off merger on antitrust concerns. *Wall Street Journal*, October 5.

Coase, Robert (1972). Durability and monopoly. *Journal of Law and Economics*, 15(1): 143–149.

Cockburn, Iain and Rebecca Henderson (1994). Racing to invest? The dynamics of competition in ethical drug discovery. *Journal of Economics and Management Strategy*, 3(3): 481–519.

Cohen, Wesley M. (2010). Fifty years of empirical studies of innovative activity and performance. In Bronwyn H. Hall and Nathan Rosenberg (Eds.), *Handbook of the Economics of Innovation* (Vol. 1) (pp. 129–213). Elsevier.

Cohen, Wesley M., Richard R. Nelson, and John P. Walsh (2004). Protecting their intellectual assets: Appropriability conditions and why US manufacturing firms patent (or not). NBER Working Paper No. 7552, February.

Cole, Bernard (2015). RISC vs CISC: What's the difference? *EE Times*, June 30. Available at https://www.eetimes.com/author.asp?section_id=36&doc_id=1327016.

Contreras, Jorge L. and Richard J. Gilbert (2015). A unified framework for RAND and other reasonable royalties. *Berkeley Technology Law Journal*, 30(2): 1451–1504.

Cournot, Augustin (1927). *Researches into the Mathematical Principles of the Theory of Wealth*. Nathaniel T. Bacon, trans. The MacMillan Co. (Original work published 1838).

Crane, Daniel A. (2011). Rethinking merger efficiencies. *Michigan Law Review*, 110(3): 347–391.

Crane, Daniel A. (2012). Search neutrality as an antitrust principle. *George Mason Law Review*, 19: 1199–1209.

Creighton, Susan A. and Jonathan M. Jacobson (2012). Twenty-five years of access denials. *Antitrust*, 27(1): 50–55.

Cukier, Kenneth (2010). Data, data, everywhere. *The Economist*, February 27.

Cunningham, Colleen, Florian Ederer, and Song Ma (2018). Killer acquisitions (August 28). Available at SSRN: https://ssrn.com/abstract=3241707.

Cusumano, Michael A. and David B. Yoffie (1998). *Competing on Internet Time: Lessons from Netscape and Its Battle with Microsoft*. Simon and Schuster.

Damanpour, Fariborz (2010). An integration of research findings of effects of firm size and market competition on product and process innovations. *British Journal of Management*, 21(4): 996–1010.

Danzon, Particia M., Andrew Epstein, and Sean Nicholson (2007). Mergers and acquisitions in the pharmaceutical and biotech industries. *Managerial and Decision Economics*, 28(4–5): 307–328.

Darby, Michael R. and Edi Karni (1973). Free competition and the optimal amount of fraud. *Journal of Law and Economics*, 16: 67–88.

Dasgupta, Partha and Eric Maskin (1987). The simple economics of research portfolios. *Economic Journal*, 97: 581–595.

Dasgupta, Partha and Joseph Stiglitz (1980). Industrial structure and the nature of innovative activity. *Economic Journal*, 90(358): 266–293.

D'Aspremont, Claude and Alexis Jacquemin (1988). Cooperative and noncooperative R&D in duopoly with spillovers. *American Economic Review*, 78(5): 1133–1137.

Delrahim, Makan (2004). Forcing firms to share the sandbox: Compulsory licensing of intellectual property rights and antitrust. Speech presented at the British Institute of International and Comparative Law, London, May 10.

Delrahim, Makan (2017). Antitrust and deregulation. Remarks as prepared for delivery at American Bar Association Antitrust Section Fall Forum, Washington, DC, November 16.

Delrahim, Makan (2018a). The "New Madison" approach to antitrust and intellectual property law. Remarks as prepared for delivery at University of Pennsylvania Law School, Philadelphia, March 16.

Delrahim, Makan (2018b). "Telegraph Road": Incentivizing innovation at the intersection of patent and antitrust law. 19th Annual Berkeley-Stanford Advanced Patent Law Institute, Palo Alto, CA, December 7.

Del Rey, Jason (2017). Amazon invested millions in the startup Nucleus—then cloned its product for the new Echo. *Vox*, May 10. Available at https://www.vox.com/2017/5/10/15602814/amazon-invested-startup-nucleus-cloned-alexa-echo-show-voice-control-touchscreen-video.

Denicolò, Vincenzo and Michele Polo (2018a). Duplicative research, mergers, and innovation. *Economics Letters*, 166(C): 56–59.

Denicolò, Vincenzo and Michele Polo (2018b). The innovation theory of harm: An appraisal. Available at SSRN: https://ssrn.com/abstract=3146731.

Deveau, Scott and Kiel Porter (2018). ComScore is exploring options including a potential sale. *Bloomberg*, January 23. Available at https://www.bloomberg.com/news/articles /2018-01-23/comscore-is-said-to-explore-options-including-a-potential-sale.

Devos, Erik, Palani-Rajan Kadapakkam and Srinivasan Krishnamurthy (2016). How do mergers create value? A comparison of taxes, market power, and efficiency improvements as explanations for synergies. *Review of Financial Studies*, 22(3): 1179–1211.

DiMasi, Joseph A., Henry G. Grabowski, and Ronald W. Hansen (2016). Innovation in the pharmaceutical industry: New estimates of R&D costs. *Journal of Health Economics*, 47: 20–33.

Doraszelski, Ulrich (2003). An R&D race with knowledge accumulation. *RAND Journal of Economics*, 34(1): 20–42.

Dryden, Benjamin R. and Shankar (Sean) Iyer (2017). Privacy fixing and predatory privacy: The intersection of big data, privacy policies, and antitrust. *Competition Policy International Antitrust Chronicle*, September.

Easterbrook, Frank H. (1992). Ignorance and antitrust. In Thomas M. Jorde and David J. Teece (Eds.), *Antitrust, Innovation, and Competitiveness*. Oxford University Press.

Elhauge, Einer (2003). Defining better monopolization standards. *Stanford Law Review*, 56: 253–344.

Elhauge, Einer (2009). Tying, bundled discounts, and the death of the single monopoly profit theory. *Harvard Law Review*, 123(2): 397–481.

Ellison, Glenn and Drew Fudenberg (2000). The neo-Luddite's lament: Excessive upgrades in the software industry. *RAND Journal of Economics*, 31(2): 253–272.

European Commission (2004). Guidelines on the assessment of horizontal mergers under the Council Regulation on the control of concentrations between undertakings (2004/C 31/03).

European Commission (2016). EU merger control and innovation. Competition Policy Brief, April.

Evans, David S., Albert L. Nichols, and Bernard Reddy (2002). The rise and fall of leaders in personal computer software. In David S. Evans (Ed.), *Microsoft, Antitrust, and the New Economy: Selected Essays* (pp. 265–285). Kluwer Academic Publishers.

Evans, David S. and Michael Noel (2005). Defining antitrust markets when firms operate two-sided platforms. *Columbia Business Law Review*, 2005(3): 667–702.

Evans, David S. and Michael Salinger (2005). Why do firms bundle and tie? Evidence from competitive markets and implications for tying law. *Yale Journal on Regulation*, 22(1): 37–89.

Evans, David S. and Richard Schmalensee (2018). Two-sided red herrings. *CPI Antitrust Chronicle*, October.

Farrell, Joseph, John Hayes, Carl Shapiro, and Theresa Sullivan (2007). Standard setting, patents, and hold-up. *Antitrust Law Journal*, 74(3): 603–70.

Farrell, Joseph and Michael L. Katz (1998). The effects of antitrust and intellectual property law on compatibility and innovation. *Antitrust Bulletin*, Fall/Winter: 609–650.

Farrell, Joseph and Michael L. Katz (2000). Innovation, rent extraction, and integration in systems markets. *Journal of Industrial Economics*, 48(4): 413–432.

Farrell, Joseph and Michael L. Katz (2005). Competition or predation? Consumer coordination, strategic pricing and price floors in network markets. *Journal of Industrial Economics*, 53(2): 203–231.

Farrell, Joseph, Janis K. Pappalardo, and Howard Shelanski (2010). Economics at the FTC: Mergers, dominant-firm conduct, and consumer behavior. *Review of Industrial Organization*, 37: 263–277.

Farrell, Joseph and Garth Saloner (1985). Standardization, compatibility, and innovation. *RAND Journal of Economics*, 16: 70–83.

Farrell, Joseph and Garth Saloner (1986). Installed base and compatibility: Innovation, product preannouncements, and predation. *American Economic Review*, 76(5): 940–55.

Farrell, Joseph and Carl Shapiro (2010). Antitrust evaluation of horizontal mergers: An economic alternative to market definition. *B. E. Journal of Theoretical Economics*, 10(1): 1–41.

Federico, Giulio, Gregor Langus, and Tommaso Valletti (2017). A simple model of mergers and innovation. *Economics Letters*, 157(C): 136–140.

Federico, Giulio, Gregor Langus, and Tommaso Valletti (2018). Horizontal mergers and product innovation: An economic framework. *International Journal of Industrial Organization*, 59: 1–23.

Federico, Giulio, Fiona Scott-Morton, and Carl Shapiro (2020). Antitrust and innovation: Welcoming and protecting disruption. Forthcoming in NBER, Innovation Policy and the Economy.

Filistrucchi, Lapo, Damien Geradin, Eric van Damme, and Pauline Affeldt (2014). Market definition in two-sided markets: Theory and practice. *Journal of Competition Law and Economics*, 10(2): 293–339.

First, Harry and Andrew I. Gavil (2006). Re-framing Windows: The durable meaning of the Microsoft antitrust litigation. *Utah Law Review*, 679(3): 679–761.

Fisher, Lawrence M. (1994). Company news: Rhone unit focuses on gene drugs. *New York Times*, November 15. Available at https://www.nytimes.com/1994/11/15/business/company-news-rhone-unit-focuses-on-gene-drugs.html.

Fox, Justin (2017). The fall, rise, and fall of creative destruction. *Bloomberg*, September 26.

Franco, April Mitchell and Darren Filson (2006). Spin-outs: Knowledge diffusion through employee mobility. *RAND Journal of Economics*, 37(4): 841–860.

Fudenberg, Drew, Richard Gilbert, Joseph Stiglitz, and Jean Tirole (1983). Preemption, leapfrogging and competition in patent races. *European Economic Review*, 22(1): 3–31.

Fuglie, Keith O. et al. (2011). Research investments and market structure in the food processing, agricultural input, and biofuel industries worldwide. US Department of Agriculture. Economic Research Report No. 130, December.

Furman, Jason (2016). Beyond antitrust: The role of competition policy in promoting inclusive growth. Speech presented at the Searle Center Conference on Antitrust Economics and Competition Policy, Chicago, September 16.

Furman, Jason and Peter Orszag (2015). A firm-level perspective on the role of rents in the rise in inequality. Paper presented at Columbia University's "A Just Society" Centennial Event in honor of Joseph Stiglitz, New York, October 16.

Galasso, Alberto and Mark Schankerman (2015). Patents and cumulative innovation: Causal evidence from the courts. *Quarterly Journal of Economics*, 130(1): 317–369.

Galasso, Alberto and Mark Schankerman (2018). Patent rights, innovation, and firm exit. *RAND Journal of Economics*, 49(1): 64–86.

Galetovic, Alexander, Stephen Haber, and Ross Levine (2015). An empirical examination of patent hold-up. *Journal of Competition Law and Economics*, 11(3): 549–578.

Gans, Joshua (2016). *The Disruption Dilemma*. MIT Press.

Garcia-Macia, Daniel, Chang-Tai Hsieh, and Peter Klenow (2017). How destructive is innovation? US Census Bureau Center for Economic Studies Paper No. CES-WP-17-04, January 1. Available at SSRN: https://ssrn.com/abstract=2896913.

Garud, Rahu, Sanjay Jain, and Arun Kumaraswamy (2002). Institutional entrepreneurship in the sponsorship of common technological standards: The case of Sun Microsystems and Java. *Academy of Management Journal*, 45(1): 196–214.

Garza, Deborah A. et al. (2007). *Antitrust Modernization Commission: Report and Recommendations*. Available at http://www.amc.gov/report_recommendation/amc_final_report.pdf.

Gavil, Andrew I. and Harry First (2014). *The Microsoft Cases: Competition Policy for the Twenty-First Century*. MIT Press.

Geradin, Damien and Dimitrios Katsifis (2018). An EU competition law analysis of online display advertising in the programmatic age. Tilburg Law & Economics Center (TILEC) working paper. Available at https://ssrn.com/abstract=3299931.

Geradin, Damien and Miguel Rato (2007). Can standard setting lead to exploitative abuse? A dissonant view on patent hold-up, royalty stacking and the meaning of FRAND. *European Competition Journal*, 3: 101–161.

Ghosh, Aloke (2001). Does operating performance really improve following corporate acquisitions? *Journal of Corporate Finance*, 7(2): 151–78.

Gilbert, Richard J. (1999). Networks, standards, and the use of market dominance: Microsoft (1995). In John E. Kwoka Jr. and Lawrence J. White (Eds.), *The Antitrust Revolution: Economics, Competition, and Policy* (3rd ed.), (pp. 409–429). Oxford University Press,

Gilbert, Richard J. (2006). Looking for Mr. Schumpeter: Where are we in the competition-innovation debate? In A. Jaffe, J. Lerner, and S. Stern (Eds.), *Innovation Policy and the Economy* (Vol. 6) (pp. 159–215). National Bureau of Economic Research.

Gilbert, Richard J. (2007). Holding innovation to an antitrust standard. *Competition Policy International*, 3(1): 3–33.

Gilbert, Richard J. (2011). Deal or no deal? Licensing negotiations in standard-setting organizations. *Antitrust Law Journal*, 77(3): 855–888.

Gilbert, Richard J. (2014). Competition policy for industry standards. In Roger Blair and D. Daniel Sokol (Eds.), *Oxford Handbook on International Antitrust Economics, Vol. 2* (pp. 554–585). Oxford University Press.

Gilbert, Richard J. (2018a). Collective rights organizations: A guide to benefits, costs and antitrust safeguards. In Jorge L. Contreras (Ed.), *The Cambridge Handbook of Technical Standardization Law: Competition, Antitrust, and Patents* (pp. 125–146). Cambridge University Press.

Gilbert, Richard J. (2018b). US Federal Trade Commission investigation of Google Search (2013). In John E. Kwoka and Lawrence White (Eds.), *The Antitrust Revolution* (7th ed.) (pp. 489–513). Oxford University Press,

Gilbert, Richard J. (2019). Competition, mergers, and R&D diversity. *Review of Industrial Organization*, 54(3): 465–484.

Gilbert, Richard and Eirik Gaar Kristiansen (2018). Licensing and innovation with imperfect contract enforcement. *Journal of Economics and Management Strategy*, 27(2): 297–314.

Gilbert, Richard J. and Hillary Greene (2015). Merging innovation into antitrust agency enforcement of the Clayton Act. *George Washington Law Review*, 83: 1919–1947.

Gilbert, Richard J. and Michael L. Katz (2001). An economist's guide to US v. Microsoft. *Journal of Economic Perspectives*, 15(2): 25–44.

Gilbert, Richard J. and David M. G. Newbery (1982). Preemptive patenting and the persistence of monopoly. *American Economic Review*, 72(3): 514–526.

Gilbert, Richard, Christian Riis, and Erlend Riis (2018). Stepwise innovation by an oligopoly. *International Journal of Industrial Organization*, 61: 413–438.

Gilbert, Richard J. and Stephen C. Sunshine (1995). Incorporating dynamic efficiency concerns in merger analysis: The use of innovation markets. *Antitrust Law Journal*, 63(2): 574–581.

Gilbert, Richard J. and Willard Tom (2001). Is innovation king at the antitrust agencies? The Intellectual Property Guidelines five years later. *Antitrust Law Journal*, 69: 43–86.

Goettler, Ronald and Brett Gordon (2011). Does AMD spur Intel to innovate more? *Journal of Political Economy*, 119 (6): 1141–1200.

Goodman, David J. and Robert A. Myers (2005). 3G cellular standards and patents. *2005 International Conference on Wireless Networks, Communications, and Mobile Computing*, 1: 415–420.

Grabowski, Henry G. and Margaret Kyle (2008). Mergers and alliances in pharmaceuticals: Effects on innovations and R&D productivity. In Klaus Gugler and Burcin Yurtoglu, (Eds.), *The Economics of Corporate Governance and Mergers*. Edward Elgar.

Greenstein, Shane and Garey Ramey (1998). Market structure, innovation and vertical product differentiation. *International Journal of Industrial Organization*, 16: 285–311.

Grindley, Peter C. and David J. Teece (1997). Managing intellectual capital: Licensing and cross-licensing in semiconductors and electronics. *California Management Review*, 39(2): 8–41.

Grove, Andrew S. (1996). *Only the Paranoid Survive: How to Exploit the Crisis Points that Challenge Every Company and Career*. Doubleday.

Gutiérrez, Germán and Thomas Philippon (2018). Declining competition and investment in the US. NBER Working Paper.

Hall, Bronwyn and Rosemarie Ham Ziedonis (2001). The patent paradox revisited: An empirical study of patenting in the U.S. Semiconductor Industry, 1979–1995. *RAND Journal of Economics*, 32(1): 101–128.

Harris, Christopher and John Vickers (1985). Perfect equilibrium in a model of a race. *Review of Economic Studies*, 52: 193–209.

Hart, Oliver D. (1983). The market mechanism as an incentive scheme. *Bell Journal of Economics*, 14(2): 366–382.

Haucap, Justus, Alexander Rasch, and Joel Stiebale (2019). How mergers affect innovation: Theory and evidence. *International Journal of Industrial Organization*, 63: 283–325.

Hawking, Stephen W. (2005). *The Theory of Everything: The Origin and Fate of the Universe*. Phoenix Books. First published under the title *The Cambridge Lectures: Life Works* (1996), Dove Audio.

Heiner, David (2012). Microsoft: A remedial success? *Antitrust Law Journal*, 78(2): 329–362.

Henderson, Rebecca (1993). Underinvestment and incompetence as responses to radical innovation: Evidence from the photolithographic alignment equipment industry. *RAND Journal of Economics*, 24(2): 248–270.

Henderson, Rebecca M. and Kim B. Clark (1990). Architectural innovation: The reconfiguration of existing product technologies and the failure of established firms. *Administrative Science Quarterly*, 35(1): 9–30.

Hermalin, Benjamin E. (1992). The effects of competition on executive behavior. *RAND Journal of Economics*, 23(3): 350–365.

Herndon, Astead W. (2019). Elizabeth Warren proposes breaking up tech giants like Amazon and Facebook. *New York Times*, March 8.

Hesse, Renata B. (2009). Section 2 remedies and US v. Microsoft: What is to be learned? *Antitrust Law Journal*, 75(3): 847–869.

Hesse, Renata B. (2014). At the intersection of antitrust and high-tech: Opportunities for constructive engagement. Remarks as prepared for the Conference on Competition and IP Policy in High-Technology Industries, Stanford, CA, January 22.

Hicks, John R. (1935). Annual survey of economic theory: The theory of monopoly. *Econometrica*, 3(1): 1–20.

Hill, Nicolas, Nancy L. Rose, and Tor Winston (2015). Economics at the Antitrust Division 2014–2015: Comcast/Time Warner Cable and Applied Materials/Tokyo Electron. *Review of Industrial Organization*, 47(4): 425–435.

Hoerner, Robert J. (1995). Innovation markets: New wine in old bottles? *Antitrust Law Journal*, 64: 49–73.

Holmstrom, Bengt (1982). Moral hazard in teams. *Bell Journal of Economics*, 13(2): 324–340.

Hombert, Johan and Adrien Matray (2018). Can innovation help U.S. manufacturing firms escape import competition from China? *Journal of Finance*, 73(5): 2003–2038.

Hovenkamp, Herbert J. (2008a). *The Antitrust Enterprise: Principles and Execution*. Harvard University Press.

Hovenkamp, Herbert J. (2008b). Schumpeterian competition and antitrust. *Competition Policy International*, 4: 273–281.

Hovenkamp, Herbert J. (2010). The Federal Trade Commission and the Sherman Act. *Florida Law Review*, 62: 1–23.

Hovenkamp, Herbert J. (2013). Implementing antitrust's welfare goals. *Fordham Law Review*, 81: 2471–2496.

Hovenkamp, Herbert J. (2017). Appraising merger efficiencies. *George Mason Law Review*, 24: 703–741.

Hunt, Robert M. (2004). Patentability, industry structure, and innovation. *Journal of Industrial Economics*, 52(3): 401–425.

Hylton, Keith N. and Haizhen Lin (2013). Innovation and optimal punishment, with antitrust applications. *Journal of Competition Law & Economics*, 10: 1–25.

Igami, Mitsuru (2017). Estimating the Innovator's Dilemma: Structural analysis of creative destruction in the hard disk drive industry, 1981–1998. *Journal of Political Economy*, 125(3): 798–847.

Igami, Mitsuru and Kosuke Uetake (2018). Mergers, innovation, and entry-exit dynamics: Consolidation of the hard disk drive industry, 1996–2016. November 22. Available at SSRN: https://ssrn.com/abstract=2585840.

Isaacson, Walter (2011). *Steve Jobs*. Simon and Schuster.

Isaacson, Walter (2014). *The Innovators: How a Group of Hackers, Geniuses, and Geeks Created the Digital Revolution*. Simon and Schuster.

Jobs, Steve (2004). Voices of innovation. *Bloomberg BusinessWeek*, October 10.

Jones, Charles I. and John C. Williams (1998). Measuring the social return to R&D. *Quarterly Journal of Economics*, 113(4): 1119–1135.

Jullien, Bruno and Yassine Lefouili (2018). Horizontal mergers and innovation. Toulouse School of Economics, Working Paper No. 18–892, May.

Katz, Michael L. and Carl Shapiro (1985). Network externalities, competition, and compatibility. *American Economic Review*, 75(3): 424–440.

Katz, Michael L., and Carl Shapiro (1986a). Product compatibility choice in a market with technological progress. *Oxford Economic Papers*, Special Issue on the New Industrial Economics, 38(1): 146–165.

Katz, Michael L. and Carl Shapiro (1986b). Technology adoption in the presence of network externalities. *Journal of Political Economy*, 94(4): 822–841.

Katz, Michael L. and Carl Shapiro (1987). Research and development rivalry with licensing or imitation. *American Economic Review*, 77(3): 402–420.

Katz, Michael L. and Carl Shapiro (1992). Product introduction with network externalities. *Journal of Industrial Economics*, 15(1): 55–83.

Katz, Michael L. and Carl Shapiro (1994). Systems competition and network effects. *Journal of Economic Perspectives*, 8(1): 93–115.

Katz, Michael L. and Howard A. Shelanski (2005). Merger policy and innovation: Must enforcement change to account for technological change? *Innovation Policy and the Economy*, 5: 109–165.

Katz, Michael L. and Howard A. Shelanski (2007a). Merger analysis and the treatment of uncertainty: Should we expect better? *Antitrust Law Journal*, 74: 537–574.

Katz, Michael L. and Howard A. Shelanski (2007b). Mergers and innovation. *Antitrust Law Journal*, 74(1): 1–85.

Kearns, David T. and David A. Nadler (1992). *Prophets in the Dark: How Xerox Reinvented Itself and Beat Back the Japanese*. Harper Collins.

Khan, Lina M. (2017). Amazon's antitrust paradox. *Yale Law Journal*, 126: 710–805.

King, Andrew (2017). The theory of disruptive innovation: Science or allegory? *Entrepreneur and Innovation Exchange*. Published online at EIX.org on October 26, 2017.

Kitch, Edmund W. (1977). The nature and function of the patent system. *Journal of Law and Economics*, 20(2): 265–290.

Klein, Joel I. (1997). Cross-licensing and antitrust law. Speech before the American Intellectual Property Law Association, May 2. Available at http://www.usdoj.gov/atr/public/speeches/1118.pdf.

Kolasky, William J. and Andrew R. Dick (2003). The merger guidelines and the integration of efficiencies into antitrust review of horizontal mergers. *Antitrust Law Journal*, 71: 207–251.

Kornfield, Rachel, J. Donohoe, E. R. Berndt, and G. C. Alexander. (2013). Promotion of prescription drugs to consumers and providers, 2001–2010. *PLOS ONE*, 8(3): 1–7.

Koyré, Alexandre (1952). An unpublished letter of Robert Hooke to Isaac Newton. *Isis*, 43(4): 312–337.

Kueng, Lorenz, Nicholas Li, and Mu-Jeung Yang (2016). The impact of emerging market competition on innovation and business strategy. NBER Working Paper No. 22840, November.

Kühn, Kai-Uwe and John Van Reenen (2009). Interoperability and market foreclosure in the European Microsoft case. In Bruce Lyons (Ed.), *Cases in European Competition Policy*. Cambridge University Press.

Kwoka, John (2008). Eliminating potential competition. *Issues in Competition Law and Policy*, 2 (ABA Section of Antitrust Law): 1437–1454.

Kwoka, John, Daniel Greenfield, and Chengyan Gu (2015). *Mergers, Merger Control, and Remedies: A Retrospective Analysis of US Policy*. MIT Press.

Kwon, Illoong (2010). R&D portfolio and market structure. *Economic Journal*, 120(543): 313–323.

Langford, Andrew (2013). gMonopoly: Does search bias warrant antitrust or regulatory intervention? *Indiana Law Journal*, 88: 1559–1592.

Lee, Robin S. (2013). Vertical integration and exclusivity in platform and two-sided markets. *American Economic Review*, 103(7): 2960–3000.

Lee, Robin S. (2014). Competing platforms. *Journal of Economics & Management Strategy*, 23(3): 507–526.

Lee, Tom and Louis L. Wilde (1980). Market structure and innovation: A reformulation. *Quarterly Journal of Economics*, 94(2): 429–436.

Lee, William F. and A. Douglas Melamed (2016). Breaking the vicious cycle of patent damages. *Cornell Law Review*, 101: 385–466.

Lemley, Mark A. and Carl Shapiro (2007). Patent hold-up and royalty stacking. *University of Texas Law Review*, 85: 1991–2049.

Lepore, Jill (2014). The disruption machine: What the gospel of innovation gets wrong. *New Yorker*, June 23.

Lerner, Joshua (1997). An empirical exploration of a technology race. *RAND Journal of Economics*, 28(2): 228–247.

Lerner, Josh and Jean Tirole (2006). A model of forum shopping. *American Economic Review*, 96(4): 1091–1113.

Letina, Igor (2016). The road not taken: Competition and the R&D portfolio. *RAND Journal of Economics*, 47(2): 433–460.

Levin, Richard C., Wesley M. Cohen, and David C. Mowery (1985). R&D appropriability, opportunity, and market structure: New evidence on some Schumpeterian hypotheses. *American Economic Review*, 75(2): 20–24.

Levin, Richard C., Alvin K. Klevorick, Richard R. Nelson, and Sidney G. Winter (1987). Appropriating the returns from industrial R&D. *Brookings Papers on Economic Activity*, 3: 783–820.

Lewis, Tracy R. (1983). Preemption, divestiture, and forward contracting in a market dominated by a single firm. *American Economic Review*, 73(5): 1092–1101.

Lipsey, R. G. and Kelvin Lancaster (1956). The general theory of second best. *Review of Economic Studies*, 24(1): 11–32.

Lohr, Steve and James Kantor (2019). A.M.D.-Intel settlement won't end their woes. *New York Times*, November 12. Available at https://www.nytimes.com/2009/11/13/technology/companies/13chip.html.

López, Ángel L. and Xavier Vives (2019). Overlapping ownership, R&D spillovers, and antitrust policy. *Journal of Political Economy*, 127(5): 2394–2437.

Loury, Glenn C. (1979). Market structure and innovation. *Quarterly Journal of Economics*, 93(3): 395–410.

Luca, Michael et al. (2015). Does Google content degrade Google search? Experimental evidence. Harvard Business School, Working Paper 16–035.

Lynn, Leonard H. (1998). The commercialization of the transistor radio in Japan: The functioning of an innovation community. *IEEE Transactions on Engineering Management*, 45(3): 220–229.

Macher, Jeffrey, Nathan H. Miller, and Matthew Osborne (2017). Finding Mr. Schumpeter: An empirical study of competition and technology adoption. Georgetown University, Working Paper, January 3.

Maddaus, Gene (2017). Nielsen sues Comscore to block new TV audience measurement service. *Variety*, September 22. Available at https://variety.com/2017/tv/news/nielsen-comscore-extended-tv-lawsuit-1202566367/.

Manne, Geoffrey A. and Joshua D. Wright (2011). Google and the limits of antitrust: The case against the case against Google. *Harvard Journal of Law and Public Policy*, 34: 1–74.

Marshall, Guillermo and Álvaro Parra (2019). Innovation and competition: The role of the product market. *International Journal of Industrial Organization*, 65: 221–247.

Maskus, Keith and Stephen A. Merrill (Eds.) (2013). *Patent Challenges for Standard-Setting in the Global Economy: Lessons from Information and Communications Technology*. National Research Council of the National Academies. National Academies Press.

Mauboussin, Michael J., Dan Callahan, and Darius Majd (2017). The incredible shrinking universe of stocks: The causes and consequences of fewer U.S. equities. Credit Suisse Report. March 22.

McCraw, Thomas K. (2012). Joseph Schumpeter on competition. *Competition Policy International*, 8: 194–221.

Melamed, Douglas A. (2006). Exclusive dealing arrangements and other exclusionary conduct—are there unifying principles? *Antitrust Law Journal*, 73: 375–412.

Merges, Robert P. and Jeffery M. Kuhn (2009). An estoppel doctrine for patented standards. *California Law Review*, 97(1): 1–50.

Merges, Robert P. and Richard R. Nelson (1990). On the complex economics of patent scope. *Columbia Law Review*, 90(4): 839–916.

Moser, Petra and Alessandra Voena (2012). Compulsory licensing: Evidence from the Trading with the Enemy Act. *American Economic Review*, 102(1): 396–427.

Motta, Massimo and Emanuele Tarantino (2017). The effect of horizontal mergers, when firms compete in prices and investments. UPF Working Paper No.1579, August 30.

Mowery, David C. (2011). Federal policy and the development of semiconductors, computer hardware, and computer software: A policy model for climate change R&D? In Rebecca M. Henderson and Richard G. Newell, (Eds.), *Accelerating Energy Innovation: Insights from Multiple Sectors* (pp. 159–188). University of Chicago Press.

Nalebuff, Barry (2004). Bundling as an entry barrier. *Quarterly Journal of Economics*, 119(1): 159–187.

Nalebuff, Barry J. and Joseph E. Stiglitz (1983). Prizes and incentives: Towards a general theory of compensation and competition. *Bell Journal of Economics*, 14(1): 21–43.

National Science Foundation (2018a). *Business Research and Development and Innovation: 2015*. National Center for Science and Engineering Statistics.

National Science Foundation (2018b). *Science and Engineering Indicators*. National Science Board.

Newman, John M. (2012). Anticompetitive product design in the new economy. *Florida State University Law Review*, 39(3): 1–54.

Nickell, Stephen J. (1996). Competition and corporate performance. *Journal of Political Economy*, 104(4): 724–746.

O'Donoghue, Ted (1998). A patentability requirement for sequential innovation. *RAND Journal of Economics*, 29(4): 654–79.

O'Donoghue, Ted, Suzanne Scotchmer, and Jacques-Francois Thisse (1998). Patent breadth, patent life, and the pace of technological progress. *Journal of Economics and Management Strategy*, 7(1): 1–32.

OECD (2018). *Oslo Manual, The Measurement of Scientific, Technological, and Innovation Activities*, 4th ed.

Ordover, Janusz A., Garth Saloner, and Steven C. Salop (1990). Equilibrium vertical foreclosure. *American Economic Review*, 80(1): 127–142.

Ordover, Janusz A. and Robert D. Willig (1981). An economic definition of predation: Pricing and product innovation. *Yale Law Journal*, 91(1): 8–53.

Ornaghi, Carmine (2009). Mergers and innovation in big pharma. *International Journal of Industrial Organization*, 27(1): 70–79.

Patterson, Mark R. (2013). Google and search-engine market power. *Harvard Journal of Law and Technology*, Occasional Paper Series: 1–24.

Petit, Nicolas (2018). Innovation competition, unilateral effects and merger control policy. International Center for Law and Economics, White Paper 2018–03.

Phillips McDougall (2016). *Agrochemical Research and Development*. March.

Pleatsikas, Christopher and David Teece (2001). The analysis of market definition and market power in the context of rapid innovation. *International Journal of Industrial Organization*, 19(5): 665–693.

Porter, Michael E. (2001). Competition and antitrust: Toward a productivity-based approach to evaluating mergers and joint ventures. *Antitrust Bulletin*, Winter: 919–958.

Posner, Richard (2001). Antitrust in the new economy. *Antitrust Law Journal*, 68: 925–943.

Ratliff, James D. and Daniel L. Rubinfeld (2010). Online advertising: Defining relevant markets. *Journal of Competition Law and Economics*, 6: 653–686.

Ratliff, James and Daniel L. Rubinfeld (2013). The use and threat of injunctions in the RAND context. *Journal of Competition Law and Economics*, 9(1): 1–22.

Ratliff, James D. and Daniel L. Rubinfeld (2014). Is there a market for organic search engine results, and can their manipulation give rise to antitrust liability? *Journal of Competition Law and Economics*, 10: 517–541.

Reinganum, Jennifer (1981). Dynamic games of innovation. *Journal of Economic Theory*, 25(1): 21–41.

Reinganum, Jennifer (1983). Uncertain innovation and the persistence of monopoly. *American Economic Review*, 73(4): 741–748.

Reinganum, Jennifer (1984). Gilbert, Richard J. and David M. G. Newbery. Uncertain innovation and the persistence of monopoly: Comment. *American Economic Review*, 74(1): 238–42.

Reinganum, Jennifer (1989). The timing of innovation: research, development, and diffusion. In R. Schmalensee and R. Willig (Eds.), *Handbook of Industrial Organization* (pp. 849–908). Elsevier Science.

Reynolds, Michael J. and Christopher Best (2012). Article 102 and innovation: The journey since Microsoft. Paper presented at the 39th Annual Fordham Conference on International Antitrust Law and Policy, September 20–21.

Rochet, Jean-Charles and Jean Tirole (2002). Cooperation among competitors: Some economics of payment card associations. *RAND Journal of Economics*, 33(4): 549–570.

Rochet, Jean-Charles and Jean Tirole (2006). Two-sided markets: A progress report. *RAND Journal of Economics*, 37(3): 645–667.

Royall, Sean M. and Adam J. DiVincenzo (2010). Evaluating mergers between potential competitors under the new Horizontal Merger Guidelines. *Antitrust*, 25(1): 33–38.

Rubinfeld, Daniel L. (2008). Maintenance of monopoly: U.S. v. Microsoft. In John E. Kwoka Jr. and Lawrence J. White (Eds.), *The Antitrust Revolution* (5th ed.) (pp. 530–557). Oxford University Press.

Rubinfeld, Daniel L. and Michal Gal (2017). Access barriers to big data. *Arizona Law Review*, 59: 339–381.

Sabety, Ted (2005). Nanotechnology innovation and the patent thicket: Which IP policies promote growth? *Albany Law Journal of Science and Technology*, 15: 477–516.

Sah, Raaj Kumar and Joseph E. Stiglitz (1987). The invariance of market innovation to the number of firms. *RAND Journal of Economics*, 18(1): 98–108.

Salant, Stephen W. (1984). Preemptive patenting and the persistence of monopoly: Comment. *American Economic Review*, 74(1): 247–250.

Salinger, Michael A. (2016). Net innovation pressure in merger analysis. Available at SSRN: https://ssrn.com/abstract=3051249.

Salinger, Michael A. and Robert J. Levinson (2015). Economics and the FTC's Google investigation. *Review of Industrial Organization*, 46: 25–57.

Salop, Steven C. (2006). Exclusionary conduct, effect on consumers, and the flawed profit-sacrifice standard. *Antitrust Law Journal*, 73: 311–374.

Salop, Steven C. and R. Craig Romaine (1999). Preserving monopoly: Economic analysis, legal standards, and Microsoft. *George Mason Law Review*, 7: 617–671.

Scharfstein, David (1988). Product-market competition and managerial slack. *RAND Journal of Economics*, 19(1): 147–155.

Scherer, F. M. (1965). Firm size, market structure, opportunity, and the output of patented inventions. *American Economic Review*, 55(5): 1097–1125.

Scherer, Frederic M. (1977). *The Economic Effects of Compulsory Patent Licensing*. New York University, Graduate School of Business Administration, Center for the Study of Financial Institutions.

Schmalensee, Richard (1999). Bill Baxter in the antitrust arena: An economist's appreciation. *Stanford Law Review*, 51(5): 1317–1332.

Schmidt, Klaus M. (1997). Managerial incentives and product market competition. *Review of Economic Studies*, 64(2): 191–213.

Schumpeter, Joseph A. (1942). *Capitalism, Socialism, and Democracy*. Harper.

Scotchmer, Suzanne (1991). Standing on the shoulders of giants: Cumulative research and the patent law. *Journal of Economic Perspectives*, 5(1): 29–41.

Scotchmer, Suzanne (2004). *Innovation and Incentives*. MIT Press.

Segal, Ilya and Michael D. Whinston (2007). Antitrust in innovative industries. *American Economic Review*, 97(5): 1703–1730.

Shapiro, Carl (1999). Exclusivity in network industries. *George Mason Law Review*, 7: 673–683.

Shapiro, Carl (2001). Navigating the patent thicket: Cross licensing, patent pools, and standard setting. In Adam Jaffe et al. (Eds.), *Innovation Policy and the Economy*. MIT Press.

Shapiro, Carl (2009). Microsoft: A remedial failure. *Antitrust Law Journal*, 75(3): 739–772.

Shapiro, Carl (2012). Competition and innovation: Did Arrow hit the bull's eye? In Josh Lerner and Scott Stern, (Eds.), *The Rate and Direction of Inventive Activity Revisited* (pp. 361–404). University of Chicago Press.

Shapiro, Carl and Hal R. Varian (1999a). The art of standards wars. *California Management Review*, 41(2): 8–32.

Shapiro, Carl and Hal R. Varian (1999b). *Information Rules: A Strategic Guide to the Network Economy*. Harvard Business School Press.

Shelanski, Howard A. and J. Gregory Sidak (2001). Antitrust divestiture in network industries. *University of Chicago Law Review*, 68(1): 1–93.

Sidak, J. Gregory and David F. Teece (2009). Dynamic competition in antitrust law. *Journal of Competition Law and Economics*, 5(4): 581–631.

Simcoe, Tim (2012). Standard setting committees: Consensus governance for shared technology platforms. *American Economic Review*, 102(1): 305–336.

Sivinski, Greg, Alex Okuliar, and Lars Kjolbye (2017). Is big data a big deal? A competition law approach to big data. *European Competition Law Journal*, 13(2–3): 199–227.

Spence, A. Michael (1984). Cost reduction, competition, and industry performance. *Econometrica*, 52(1): 101–122.

Srinivason, Dina (2019). The antitrust case against Facebook: A monopolist's journey towards pervasive surveillance in spite of consumers' preference for privacy. *Berkeley Business Law Journal*, 16(1): 39–101.

Stewart, Marion B. (1983). Noncooperative oligopoly and preemptive innovation without winner-take-all. *Quarterly Journal of Economics*, 98(4): 681–694.

Stigler, George J. (1963). United States v. Loew's Inc.: A note on block-booking. *Supreme Court Review*, 1963: 152–157.

Stiglitz, Joseph E. (2017). Towards a broader view of competition policy. Roosevelt Institute, Working Paper, June.

Stone, Brad (2013). *The Everything Store: Jeff Bezos and the Age of Amazon*. Little Brown & Co.

Stucke, Maurice E. and Ariel Ezrachi (2016). When competition fails to optimize quality: A look at search engines. *Yale Journal of Law and Technology*, 18(1).

Sutton, John (1998). *Technology and Market Structure*. MIT Press.

Swedin, Eric G. (2009). Why OS/2 failed: Business mistakes compounded by memory prices. *Mountain Plains Journal of Business and Economics, Opinions and Experiences*, 10: 29–38.

Syverson, Chad (2011). What determines productivity? *Journal of Economic Literature*, 49(2): 326–365.

Tom, Willard (2001). The 1975 Xerox consent decree: Ancient artifacts and current tensions. *Antitrust Law Journal*, 68(3): 967–990.

Tripsas, Mary and Giovanni Gavetti (2000). Capabilities, cognition, and inertia: Evidence from digital imaging. *Strategic Management Journal*, 21(10–11): 1147–1161.

Tushman, Michael L. and Philip Anderson (1986). Technological discontinuities and organizational environments. *Administrative Science Quarterly*, 31(3): 439–465.

US Census Bureau (2015). *Business R&D and Innovation Survey*.

US Department of Justice (1968). 1968 Merger Guidelines. Available at https://www.justice.gov/archives/atr/1968-merger-guidelines.

US Department of Justice and Federal Trade Commission (various years). *Hart-Scott-Rodino Annual Report*.

US Department of Justice and Federal Trade Commission (2000). *Antitrust Guidelines for Collaborations among Competitors*. April.

US Department of Justice and Federal Trade Commission (2010). *Horizontal Merger Guidelines*. August 19.

US Department of Justice and Federal Trade Commission (2017). *Antitrust Guidelines for the Licensing of Intellectual Property*. January 12.

US Federal Trade Commission (2013a). Google Press Conference Opening Remarks of Federal Trade Commission Chairman Jon Leibowitz as Prepared for Delivery, January 3.

US Federal Trade Commission (2013b). Statement of the Federal Trade Commission Regarding Google's Search Practices, In the Matter of Google Inc. FTC File Number 111–0163, January 3.

US Federal Trade Commission (2017). *The FTC's Merger Remedies 2006–2012: A Report of the Bureaus of Competition and Economics*. January.

Van Reenen, John (2011). Does competition raise productivity through improving management quality? *International Journal of Industrial Organization*, 29(3): 306–316.

Vickers, John (1986). The evolution of market structure when there is a sequence of innovations. *Journal of Industrial Economics*, 35(1): 1–12.

Vickers, John (2005). Abuse of market power. *Economic Journal*, 115(504): F244–F261.

Vickers, John (2010). Competition policy and property rights. *Economic Journal*, 120: 375–392.

Visnji, Margaret (2019). Pharma industry merger and acquisition analysis, 1995 to 2015. *Revenues and Profits*, February 11. Available at https://revenuesandprofits.com/pharma-industry-merger-and-acquisition-analysis-1995-2015/.

Vives, Xavier (2008). Innovation and competitive pressure. *Journal of Industrial Economics*, 41(1): 419–469.

Watzinger, Martin, Thomas A. Fackler, Markus Nagler, and Monika Schnitzer (2017). How antitrust enforcement can spur innovation: Bell Labs and the 1956 Consent Decree. CESIFO Working Paper No. 6351, February.

Wells, Georgia (2019). Facebook to pull controversial Onavo app. *Wall Street Journal,* February 22.

Werden, Gregory J. (2006). The "no economic sense" test for exclusionary conduct. *Antitrust Law Journal,* 73: 413–433.

Wessner, Charles W. (Ed.) (2001). *Capitalizing on New Needs and New Opportunities: Government–Industry Partnerships in Biotechnology and Information Technologies.* Board on Science, Technology, and Economic Policy. National Research Council. National Academies Press.

Weyl, Glen E. and Alexander White (2014). Let the right "one" win: Policy lessons from the new economics of platforms. Coase-Sandor Working Paper Series in Law and Economics, No. 709.

Whinston, Michael D. (1990). Tying, foreclosure, and exclusion. *American Economic Review,* 80(4): 837–859.

Whinston, Michael D. (2001). Exclusivity and tying in US v. Microsoft: What we know, and don't know. *Journal of Economic Perspectives,* 15(2): 63–80.

Williamson, Oliver E. (1979). Transactions-cost economics: The governance of contractual relations. *Journal of Law and Economics,* 22(2): 233–262.

Williamson, Oliver E. (1983). *Markets and Hierarchies: Analysis and Antitrust Implications.* Free Press.

Wollman, Thomas (2019). Stealth consolidation: Evidence from an amendment to the Hart-Scott-Rodino Act. *American Economic Review: Insights,* 1(1): 77–94.

Wu, Tim (2018a). After consumer welfare, now what? The "protection of competition" standard in practice. *Journal of the Competition Policy International,* April.

Wu, Tim (2018b). The curse of bigness: Antitrust in the new gilded age. Columbia Global Reports.

Xu, Rui and Kaiji Gong (2017). Does import competition induce R&D reallocation? Evidence from the U.S. International Monetary Fund Working Paper 17/253.

## List of Cited Cases and Press Releases

### US cases

*Addamax v. Open Software Foundation.* US District Court for the District of Massachusetts (May 19, 1995).

*Addamax v. Open Software Foundation.* US Court of Appeals for the First Circuit (December 4, 1998).

*Allied Orthopedic Appliances, Inc. v. Tyco Health Care Group LP.* US Court of Appeals, Ninth Circuit (January 6, 2010).

*Allied Tube & Conduit Corp. v. Indian Head.* US Supreme Court (June 13, 1988).

*American Society of Mechanical Engineers v. Hydrolevel Corp.* US Supreme Court (May 17, 1982).

*Aspen Skiing Co. v. Aspen Highlands Skiing Corp.,* US Supreme Court (June 19, 1985).

*Berkey Photo v. Eastman Kodak Co.* US Court of Appeals for the Second Circuit (June 25, 1979).

*Burton v. BMW AG.* 17-cv-04314, US District Court, Northern District of California (July 28, 2017).

*California Computer Prods. v. IBM Corp.* US Court of Appeals, Ninth Circuit (June 21, 1979).

*Commonwealth Scientific & Industrial Research Organisation v. Cisco Systems.* US Court of Appeals for the Federal Circuit (December 3, 2015).

*C.R. Bard, Inc. v. M3 Systems, Inc.* US Court of Appeals for the Federal Circuit (September 30, 1998).

*Ericsson, Inc. v. D-Link Systems, Inc.* US Court of Appeals for the Federal Circuit (December 4, 2014).

*Golden Bridge Tech., Inc. v. Nokia, Inc.* US District Court for the Eastern District of Texas (September 10, 2007).

*Golden Bridge Tech., Inc. v. Motorola, Inc.* US Court of Appeals for the Fifth Circuit (October 23, 2008).

*Golden Gate Pharmacy Services, Inc. v. Pfizer, Inc.* US District Court for the Northern District of California (December 2, 2009).

*Golden Gate Pharmacy Services, Inc. v. Pfizer, Inc.* US Court of Appeals for the Ninth Circuit (May 19, 2011).

*GSI Tech., Inc. v. Cypress Semiconductor Corp.* US Dist. LEXIS 9378 (January 27, 2015).

*Illinois Tool Works Inc. v. Independent Ink, Inc.* US Supreme Court (March 1, 2006).

*In re Apple iPod iTunes Antitrust Litigation.* US District Court for the Northern District of California (May 19, 2011).

*In re Apple iPod iTunes Antitrust Litigation.* Verdict Form Re Genuine Product Improvement (N.D. Cal. 2014).

*In re Innovatio IP Ventures, LLC Patent Litig.* US District Court for the Northern District of Illinois (October 3, 2013).

*Kaufman v. BMW.* 17-cv-05440, US District Court of New Jersey (July 25, 2017).

*Lorain Journal Co. v. United States,* US Supreme Court (December 11, 1951).

*Massachusetts v. Microsoft Corp.* US Court of Appeals for the District of Columbia Circuit (June 30, 2004).

*Memorex Corp. v. IBM Corp.* US Court of Appeals, Ninth Circuit (November 18, 1980).

*Microsoft Corporation v. Motorola, Inc., et al.* US District Court for the Western District of Washington (April 25, 2013).

*New York ex. rel. v. Microsoft.* California Group's Report on Remedial Effectiveness, Civil Action 98–1233 (CKK) (D.D.C. September 11, 2007).

*New York v. Actavis PLC.* US Court of Appeals for the Second Circuit (May 22, 2015).

*Ohio et al. v. American Express Co. et al.* US Supreme Court (June 25, 2018).

*Otter Tail Power Co. v. United States.* US Supreme Court (February 22, 1973).

*Rambus Inc. v. FTC.* US Court of Appeals for the District of Columbia (April 22, 2008).

*SCM Corp. v. Xerox Corp.* US District Court for the District of Connecticut (December 29, 1978).

*SCM Corp. v. Xerox Corp.* US Court of Appeals for the Second Circuit (March 12, 1981).

*Transamerica Computer Co. v. IBM Corp.* US Court of Appeals, Ninth Circuit (February 15, 1983).

*US et al. v. Dow Chemical Corporation and E.I. DuPont de Nemours and Company.* Competitive Impact Statement (June 15, 2017).

US Federal Trade Commission. *In re Baxter Int'l Inc.*, 123 F.T.C. 904 (1997).

US Federal Trade Commission. *In re Ciba-Geigy Ltd.*, 123 F.T.C. 842 (1997).

US Federal Trade Commission. *In re Dow Chem. Co.*, 118 F.T.C. 730 (1994).

US Federal Trade Commission. *In re Eli Lilly & Co.*, 95 F.T.C. 538 (1980).

US Federal Trade Commission. *In re Institut Merieux S.A.*, 113 F.T.C. 742 (1990).

US Federal Trade Commission. *In re Roche Holding Ltd.*, 113 F.T.C. 1086 (1990).

US Federal Trade Commission. *In the Matter of Bayer AG and Aventis S.A.*, Docket No. C-4049, Analysis of Proposed Consent Order to Aid Public Comment (May 30, 2002).

US Federal Trade Commission. *In the Matter of Ciba-Geigy, Chiron, Sandoz and Novartis*, Docket No. C-3725, *Analysis of Proposed Consent Order to Aid Public Comment* (December 17, 1996).

US Federal Trade Commission. *In the Matter of Dell Computer Corporation*, Complaint (May 20, 1996).

US Federal Trade Commission. *In the Matter of Dell Computer Corporation*, Statement of the Commission (June 17, 1996).

US Federal Trade Commission. *In the Matter of Flow International Corp.*, File No. 081–0079, Complaint (August 15, 2008).

US Federal Trade Commission. *In the Matter of Flow International Corp.*, Analysis of the Agreement Containing Consent Order to Aid Public Comment File No. 081–0079.

US Federal Trade Commission. *In the Matter of Glaxo Wellcome and SmithKine Beecham*, Analysis of Proposed Consent Order to Aid Public Comment (2000).

US Federal Trade Commission. *In the Matter of Intel Corporation*, Docket No. 9341, Complaint (December 16, 2009).

US Federal Trade Commission. *In the Matter of Intel Corporation*, Docket No. 9341, Analysis of Proposed Consent Order to Aid Public Comment (August 4, 2010).

US Federal Trade Commission. *In the Matter of Intel Corporation*, Docket No. 9341, Decision and Order (October 29, 2010).

US Federal Trade Commission. *In the Matter of Medtronic, Inc. and Covidien plc*, Docket No. C-4503, Decision and Order (January 21, 2015).

US Federal Trade Commission. *In the Matter of Negotiated Data Solutions LLC*, Complaint, (September 23, 2008).

US Federal Trade Commission. *In the Matter of Nielsen Holdings N.V. and Arbitron Inc.*, File No. 131 0058, Analysis of Agreement Containing Consent Order to Aid Public Comment (September 20, 2013).

US Federal Trade Commission. *In the Matter of Novartis and AstraZeneca*, Docket No. C-3979, Complaint (November 1, 2000).

US Federal Trade Commission. *In the Matter of Novartis, AG and GlaxoSmithKline plc*, Docket No. C-4510, Complaint (February 20, 2015).

US Federal Trade Commission. *In the Matter of Novartis, AG and GlaxoSmithKline plc*, Analysis of Agreement Containing Consent Orders to Aid Public Comment, File No. 141–0141 (February 23, 2015).

US Federal Trade Commission. *In the Matter of Pfizer, Inc. and Hospira, Inc.*, Docket No. C-4537, Complaint (August 21, 2015).

US Federal Trade Commission. *In the Matter of Rambus Inc.*, Docket No. 9302, Opinion of the Commission (August 2, 2006).

US Federal Trade Commission. *In the Matter of Robert Bosch Gmbh*, Complaint (November 26, 2012).

US Federal Trade Commission. *In the Matter of Thoratec Corporation and HeartWare International, Inc.*, Docket No. 9339, Complaint (July 28, 2009).

*US Federal Trade Commission v. Steris Corporation*. US District Court for the N.D. Ohio (September 25, 2015).

*US v. 3D Systems Corporation and DTM Corporation*. Complaint, US District Court for the District of Columbia (June 6, 2001).

*US v. 3D Systems Corporation and DTM Corporation*. Competitive Impact Statement, US District Court for the District of Columbia (September 4, 2001).

*US v. 3D Systems Corporation and DTM Corporation*. Final Judgment, US District Court for the District of Columbia (April 17, 2002).

*US v. Automobile Manufacturers Ass'n*. 307 F. Supp. 617 (C.D. Cal. 1969), aff'd in part and appeal dismissed in part; *City of New York v. US*, US Supreme Court (March 16, 1970).

*US v. E. I. DuPont de Nemours*. US Supreme Court (June 3, 1957).

*US v. General Motors and ZF Friedrichshafen, AG, et al.* Civil Action 93–530, Complaint (November 16, 1993).

*US v. Google and ITA Software*. Complaint, US District Court for the District of Columbia, Case: 1:11-cv-00688 (April 8, 2011).

*US v. Heraeus Electro-Nite Co., LLC*. US District Court for the District of Columbia, Complaint (January 2, 2014).

*US v. Heraeus Electro-Nite Co., LLC*. US District Court for the District of Columbia, Final Judgment (April 7, 2014).

*US v. International Business Machines*. Final Judgment, US District Court for the Southern District of New York, Civil Action No. 72–344 (1956).

*US v. Lockheed Martin Corp. and Northrop Grumman Corp.* Complaint (March 23, 1998).

*US v. Microsoft*. US District Court for the District of Columbia, Civil Action No. 98–1232 (TPJ), Complaint (May 18, 1998).

*US et al. v. Microsoft*. US District Court for the District of Columbia, Civil Action No. 98–1232 (TPJ), Findings of Fact (November 5, 1999).

*US et al. v. Microsoft*. US District Court for the District of Columbia, Civil Action No. 98–1232 (TPJ), Conclusions of Law (April 3, 2000).

*US et al. v. Microsoft*. US District Court for the District of Columbia, Civil Action No. 98–1232 (TPJ), Declaration of Carl Shapiro (April 28, 2000).

*US et al. v. Microsoft*. US District Court for the District of Columbia, Civil Action No. 98–1232 (TPJ), Declaration of Paul M. Romer (April 27, 2000).

*US v. Microsoft*. Court of Appeals, for the District of Columbia Circuit (June 28, 2001).

*US et al. v. Microsoft*. Final Judgment, US District Court for the District of Columbia (November 12, 2002).

*US et al. v. Microsoft*. Joint Status Report on Microsoft's Compliance with the Final Judgments (US D.D.C) Civil Action No. 98–1232 (April 27, 2011).

*US v. Philadelphia National Bank*. US Supreme Court (June 17, 1963).

*US v. Thomson Corporation and Reuters Group plc*. Complaint (February 19, 2008).

*Verizon Communications v. Law Offices of Curtis V. Trinko*. US Supreme Court (January 13, 2004).

*Xerox Corporation*. 86 F.T.C. 364 (1975).

## US press releases

US Department of Justice. Statement of the Department of Justice Antitrust Division on its decision to close its investigation of XM Satellite Radio Holdings Inc.'s merger with Sirius Satellite Radio Inc. March 24, 2008.

US Department of Justice. Justice Department reaches settlement with Microsemi Corp. August 20, 2009.

US Department of Justice. Statement of the Department of Justice Antitrust Division on its decision to close its investigation of the internet search and paid search advertising agreement between Microsoft Corporation and Yahoo! Inc. February 18, 2010.

US Department of Justice. Department of Justice Antitrust Division statement on the closing of its investigation of the T-Mobile/MetroPCS merger. March 12, 2013.

US Department of Justice. Lam Research Corp. and KLA-Tencor Corp. abandon merger plans. October 5, 2016.

US Federal Trade Commission. Glaxo plc. June 20, 1995.

US Federal Trade Commission. Statement of Chairman Robert Pitofsky and Commissioners Janet D. Steiger, Roscoe B. Starek III, and Christine A. Varney in the matter of the Boeing Company/McDonnell Douglas Corporation. July 1, 1997.

US Federal Trade Commission. FTC seeks to block Cytyc Corp.'s acquisition of Digene Corp. June 24, 2002.

US Federal Trade Commission. Statement of Chairman Timothy J. Muris in the matter of Genzyme Corporation/Novazyme Pharmaceuticals, Inc. January 13, 2004.

US Federal Trade Commission. Statement of the Commission concerning Google/AdMob. File no. 101–0031, May 21, 2010.

US Federal Trade Commission. Statement of the Federal Trade Commission concerning Western Digital Corporation/Viviti Technologies Ltd. and Seagate Technology LLC/Hard Disk Drive Assets of Samsung Electronics Co. Ltd. May 9, 2013.

US Federal Trade Commission. FTC accepts proposed consent order in Broadcom Limited's $5.9 billion acquisition of Brocade Communications Systems, Inc. July 3, 2017.

## Cited European Cases and Press Releases

### European cases

European Commission. *Astra Zeneca/Novartis*, Case No. COMP/M.1806 (July 26, 2000).

European Commission. *Bayer/Aventis Crop Science*, Case No. COMP/M.2547 (April 17, 2002).

European Commission. Commission Decision relating to a proceeding pursuant to Article 82 of the EC Treaty and Article 54 of the EEA Agreement against Microsoft Corporation (Case COMP/C-3/37.792—*Microsoft*) (May 24, 2004).

European Commission. Decision, Case COMP/C-3/39.530—*Microsoft (tying)* (December 16, 2009).

European Commission. *Dow/DuPont*, Case No. COMP/M.7932 (March 27, 2017).

European Commission. *General Electric/Alstom*, Case No. COMP/M.7278 (September 8, 2015).

European Commission. *Glaxo-Wellcome*, IV/M.555 (February 28, 1995).

European Commission. *Glaxo Wellcome/SmithKline Beecham*, Case No. COMP/M.1486 (August 5, 2000).

European Commission. *Google Android*, Case AT.40099, Commission Decision (July 18, 2018).

European Commission. *Google Search (Shopping)*, Case AT.39740, Decision (June 27, 2017).

European Commission. *Google Search (Shopping)*, Case AT.39740, Summary of Commission Decision (June 27, 2017).

European Commission. *IMS Health GmbH & Co. OHG v. NDC Health GmbH &Co KG*, Case C-418/01, Judgment (April 29, 2004).

European Commission. *Magill TV Guide/ITP, BBC and RTE*, OJ L78/43, Decision (December 21, 1988).

European Commission. *Medtronic/Covidien*, Case No. COMP/M.7326 (November 28, 2014).

European Commission. *Microsoft/LinkedIn*, Case M.8124 (December 6, 2016).

European Commission. *Pasteur Mérieux-Merck*, IV/34.776 (October 6, 1994).

European Commission. *Pfizer/Hospira*, Case No. COMP/M.7559 (April 8, 2015).

European Commission. *Sea Containers Sealink/Stena*, Decision, OJ 1994 L15/8 (December 21, 1993).

European Commission. *Western Digital Irland/Viviti Technologies*, Case No. COMP/M.6203 (November 23, 2011).

Judgment of the Court of First Instance. *Microsoft v. Commission*, Case T-201/04 (September 17, 2007).

Official Journal of the European Communities. Council Directive on the legal protection of computer programs, 91/250/EEC (May 14, 1991).

UK Office of Fair Trading. Anticipated acquisition by Facebook Inc of Instagram Inc., ME/5525/12, (August 14, 2012).

### European press releases

European Commission. Antitrust: Commission fines Google €1.49 billion for abusive practices in online advertising. Brussels. March 20, 2019.

European Commission. Antitrust: Commission fines Google €2.42 billion for abusing dominance as search engine by giving illegal advantage to own comparison shopping service—Factsheet. Brussels. June 27, 2017.

European Commission. Antitrust: Commission fines Google €4.34 billion for illegal practices regarding Android mobile devices to strengthen dominance of Google's search engine. Brussels. July 18, 2018.

European Commission. Antitrust: Commission fines truck producers € 2.93 billion for participating in a cartel. Brussels. July 19, 2016.

European Commission. Antitrust: Commission imposes € 899 million penalty on Microsoft for non-compliance with March 2004 decision. IP/08/318. February 27, 2008.

European Commission. Antitrust: Commission imposes fine of €1.06 bn on Intel for abuse of dominant position; orders Intel to cease illegal practices. Brussels. May 13, 2009.

European Commission. Antitrust: Commission initiates formal investigations against Microsoft in two cases of suspected abuse of dominant market position. Memo/08/19. January 14, 2008.

European Commission. Antitrust: Commission market tests Microsoft's proposal to ensure consumer choice of web browsers; welcomes further improvements in field of interoperability. Memo/09/439. October 7, 2009.

European Commission. Commission examines the impact of Windows 2000 on competition. February 10, 2000.

European Commission. Mergers: Commission approves acquisition of Hospira by Pfizer, subject to conditions. August 4, 2015.

European Commission. Mergers: Commission clears merger between Dow and DuPont, subject to conditions. March 27, 2017.

Court of Justice of the European Union. The Court of Justice sets aside the judgment of the General Court which had upheld the fine of €1.06 billion imposed on Intel by the Commission for abuse of a dominant position. September 6, 2017.

# Index

Printed in the United States
by Baker & Taylor Publisher Services